Rafaela Danielli Nicola

Influence of temperature on Caiman crocodilus yacare in the Pantanal

Rafaela Danielli Nicola

Influence of temperature on Caiman crocodilus yacare in the Pantanal

Temperature fluctuations and ecology of the species

ScienciaScripts

Imprint

Cover image: www.ingimage.com

This book is a translation from the original published under ISBN 978-613-9-67803-7.

Publisher:
Sciencia Scripts
is a trademark of
Dodo Books Indian Ocean Ltd. and OmniScriptum S.R.L publishing group

120 High Road, East Finchley, London, N2 9ED, United Kingdom
Str. Armeneasca 28/1, office 1, Chisinau MD-2012, Republic of Moldova, Europe
Printed at: see last page
ISBN: 978-620-8-17966-3

INDEX

THANKS

To Prof. Dr. Eliézer José Marques (UFMS), my advisor.

To CNPq for the master's scholarship.

To Prof. Dr. Konradin Metze (UNICAMP), for his great help in various stages of the development of this work, in the histological identification of the gonads, statistical analysis of the results and valuable suggestions.

To Prof. Masao Uetanabaro (UFMS) for his support.

To Prof. Paulo Robson de Souza (UFMS) and Màrio Moreira da Silva (UNICAMP) for their contributions to the photographic documentation.

To Prof. Dr. Guilherme Mourao (EMBRAPA) for his attention and help with the sexing of the chicks during my stay in Corumbà.

To Paulo Roberto Manzani (Department of Zoology, UNICAMP) for his interest and help during the process of choosing the method for sexing the cubs.

To Prof. Dr. Augusto Shinya Abe (UNESP, Rio Claro) for his suggestions.

To Prof. Dr. Ester Nicola, Coordinator of the LASER Laboratory of the Experimental Medicine and Surgery Center (NMCE/UNICAMP), for making it possible to use the laboratory and for helping with the slides and photographic documentation.

To Edmyr Rosa Reis and Maria Dionisio de Oliveira, technicians at the NMCE/UNICAMP LASER Laboratory, for their help during the laboratory processes for the sexual identification of the sample.

To Josué Raizer, for his suggestions on data processing and statistical assistance.

To Prof. Dr. Luciano Verdade (ESALQ) for his valuable criticism and suggestions.

To Prof. Dr. Jorge Nicola for his invaluable suggestions and discussions during all stages of the work.

To my parents, Jorge and Ester Nicola, for always being there, helping and encouraging in every possible way. For their patience and trust.

To my brother Alessandro, for his affection.

To the professors of the Zoology Department.

To the coordinators of the "Master's Degree in Ecology and Conservation" and to Chiquinho.

To the interns who took part in the field and laboratory work at the beginning of the project, especially Emilene, Maristela and Alexandre.

To Vanderlei Bertho Jr. for his support and company in the early stages of the project.

To my colleague and friend Genimar Rebouças Juliao, who was with me throughout my work and who always encouraged me. For her company during the countless "visits" to the caimans.

To Marcos Rogério Dorneles for his affection, understanding and infinite patience. For the many buckets of water and help with the laboratorial stages and the final grammar check.

To Prof. Celso Benites, responsible for the UFMS Psiculture Station, and to Telma Bazzano S. Carvalho, responsible for the UFMS vivarium, for providing space to keep the chicks.

The other people who accompanied and contributed in various ways during the years of this master's degree.

INFLUENCE OF INCUBATION TEMPERATURE ON PANTANAL JACAREE CHILDREN, *CAIMAN CROCODILUS YACARE*, IN NATURAL AND CONTROLLED CONDITIONS.

Summary:

In this work we studied the influence of temperature on the incubation of eggs of the marsh caiman (*Caiman crocodilus yacare*), collected in the Pantanal of Mato Grosso do Sul, Miranda and Abobral sub-regions. This species has a temperature-dependent sex determination mechanism (TSD), with females at low temperatures and males at high temperatures. Using thermometers installed at different depths in a forest nest with 31 eggs, located on the bank of the Miranda River, we found that the temperature inside the chamber (average of 31.89+1.48° C) was significantly higher than the outside temperature (average of 28.65±2.53oC), and all the chicks sexed after hatching (N=24) were females. We developed a suitable incubation system to verify the influence of thermal fluctuations on the sex determination of the species. In the laboratory, we subjected eggs that were no more than 5 days old to 5 different ranges of daily thermal fluctuations, with an average of 30oc. The sex of the chicks was determined by histological analysis of the gonads (N=180) and only females were born in all treatments, indicating that these daily temperature fluctuations do not interfere with sex determination. The average number of eggs per spawning was 25.93+5.37 (N=17) and the egg sizes were similar between spawnings. The chicks incubated in the laboratory were equivalent in size to those incubated under natural conditions.

Keywords: wetland caiman, *Caiman crocodilus yacare*, incubation temperature, TSD, sex determination, Pantanal sul-mato-grossense.

General Introduction

The question of sexual definition in different species has aroused the interest of several researchers in recent decades. According to Bull (1983), in some species, sex is determined after conception, depending on the immediate environment, and may be associated with an environment or condition as a formative factor for males or females, a process known as environmental sex differentiation (ESD). In these cases, the genotype of the individual has little influence on the sex of the offspring. In mammals and birds, sex is determined through genetic mechanisms (Pough, 1998). The dichotomy between the two mechanisms mentioned above does not always occur, since an organism can have a certain genotypic condition that will only be evident under the action of a certain environmental condition

(Gutzke, 1987).

In reptiles, both mechanisms can be at work: in turtles and lizards there are both species whose sex definition is due to environmental factors and species with genetic sex determination (Charnier, 1966; Yntema & Mrsovsky, 1980 and Mrosovsky *et al.*, 1984). The two mechanisms can coexist within and between phylogenetic lineages (Lang and Andrews, 1994). In crocodilians, the mechanism of sex determination through the environment occurs in all species studied to date (Ferguson & Joanen, 1982; Hutton, 1987; Webb *et al.*, 1987; Lang *et al.*, 1989 and Deeming & Ferguson, 1989) and the absence of sex chromosome heteromorphism observed by Cohen & Gans, (1970) indicates that the ESD mechanism is universal among this Order.

The evolutionary importance of environmental sex determination is still not well understood and there are controversies. Some factors seem to indicate that ESD would be an ancestral trait to genetic determination, but other evidence does not support this hypothesis, so a clear relationship between the mechanisms of sex determination through the environment and genetic load has not yet been verified (Wibbels, 1994). Bull & Charnov (1989) suggested that evolutionarily, environmental sex determination was favored in species where the environment had a different influence on the *fitness* (measured by fertility) of the male or female individual, preventing the birth of poorly fertile individuals.

Most of the studies carried out on species with ESD have confirmed the importance of temperature during the incubation period as the main factor (TSD, Temperature sex dependence). Other physical characteristics such as humidity and solar radiation have also been mentioned in various studies, but their roles in sexual differentiation are not yet well defined.

There are several species of crocodilians distributed in regions with tropical and temperate climates. The Order Crocodylia includes three sub-families: the Alligatorinae, Crocodylinae and Gavialinae (Groombridge, 1987).

Class: Reptilia

Subclass: Archosauria

Order: Crocodylia

Family: Crocodylidae

Subfamily: Alligatorinae

Genus: *Alligator* (2 species)

Genus: *Caiman* (2 species)

Genus: *Paleosuchus* (2 species)

Genus: *Melanosuchus* (1 species)

Subfamily: Crocodylinae

Genus: *Crocodylus* (12 species)

Genus: *Osteolaemus* (1 species)

Genus: *Tomistoma* (1 species)

Subfamily: Gavialinae

Genus: *Gavialis* (1 species)

The marsh caiman, *Caiman crocodilus yacare* (Daudin, 1802) is distributed in the central region of southwestern South America, covering Bolivia, Paraguay, Argentina and Brazil. In Brazil, *C. c. yacare* is frequently found in the states of Mato Grosso and Mato Grosso do Sul, inhabiting the Pantanal region (Groombridge, 1987; Brazaits *et al.*, 1990 and Marques & Monteiro, 1995). This region is characterized as a depression in the Upper Paraguay river basin, covering an area of 136,700 km^2 , subject to periodic flooding and a semi-arid climate. Average annual temperatures range from 23 to 26°C, with very hot summers reaching absolute maximums of 40 to 42°C (Boggiani & Coimbra, 1996; Adâmoli, 1986; Cadavid Garcia, 1984 and RADAMBRASIL1982a,b).

In the Pantanal, *C. c. yacare* is associated with aquatic formations, and this distribution is influenced by the local seasonal cycles of flood and drought (Marques & Monteiro, 1995). Nesting occurs during the rainy season, peaking in mid-January. Caimans build nests in the form of mounds, using organic matter, and the nesting sites are generally close to bodies of water, on floating vegetation, in riparian forests or isolated patches of vegetation (Campos & Magnuson, 1995; Crawshaw, 1987 and Marques & Monteiro, 1995).

The nesting ecology of this species has been studied by a number of researchers (e.g. Crawshaw, 1987; Campos, 1993; Campos & Magnuson, 1995 and Max *et al.*, 1997). In this study, we carried out experiments with the incubation of marsh caiman eggs in natural and controlled conditions, in an attempt to obtain information that could contribute to understanding the mechanisms that influence the species' sex ratio, as well as aspects of nesting related to reproductive investment.

This study enabled us to produce four papers, presented here, which were a result of the way our experiment was developed.

Motivated by the work of Georges *et al.* (1994), we developed a thermo-controlled incubator system and software to provide the appropriate thermal conditions for the model proposed by the authors and applied to *Caretta caretta* turtles, with the aim of testing it on *Caiman crocodilus yacare*.

Once we had developed this incubation system, presented under the title "Thermo-controlled incubators for experimental verification of models relating embryonic development and incubation temperature", on pages 24 to 33, we moved on to the material collection stage. These collections were made in the Pantanal sub-regions of Miranda and Abobral (*sensu* Adàmoli, 1986).

Simultaneously with the collection of eggs for the laboratory experiment, we selected a nest in the field and, through periodic temperature measurements in various positions, evaluated the degree of thermal insulation provided by the material that made up the nest. This work is presented on pages 8 to 23, under the title "Incubation temperature of a *Caiman crocodilus yacare* nest under natural conditions".

In the paper entitled "Effect of incubation temperature, under controlled conditions, on the sex ratio of the marsh *caiman, Caiman crocodilus yacare*.", pages 34 to 57, we applied the model in question (Georges *et al.* 1994) to the marsh caiman and evaluated the results obtained.

In order to investigate some aspects related to reproductive investment, we took standardized measurements on eggs and newly hatched offspring from the different broods that made up our sample. The results of this study are presented in the paper "Variations in the size of nests, eggs and young of the marsh caiman *Caiman crocodilus yacare*, incubated under natural and controlled conditions", pages 58 to 72.

With this study, we hope to contribute information to the existing body of data on the mechanisms of sex determination and the relationships between litters and offspring of the marsh caiman.

INCUBATION TEMPERATURE OF A *CAIMAN CROCODILUS YACARE* NEST UNDER NATURAL CONDITIONS.

Rafaela Danielli Nicola[1] , Eliézer José Marques[21]

[1]Master's Degree Course in Ecology and Conservation/CCBS, Federal University of Mato Grosso do Sul, P.O. Box 649, CEP 79070-900, Campo Grande, MS.

[2]Department of Biology/CCBS, Federal University of Mato Grosso do Sul.

Summary:

We measured the incubation temperature of a natural nest of *Caiman crocodilus yacare* located in the riparian forest of the Miranda River in the Pantanal of Mato Grosso do Sul in the Abobral region. Temperatures were monitored by taking hourly readings during the 38-day incubation period using three thermometers inserted inside the laying chamber and an external thermometer fixed just above the nest. The average temperature found for the whole period, inside the chamber, was 31.89±1.48° C and for the external thermometer 28.65±2.53° C. The results indicated that the temperatures inside the nest were higher and differed from the outside temperature (Anova $F3_{.2271=337}.56$ $p<0.001$) throughout the period, indicating the existence of significant internal heating. The maintenance of the internal temperature during the night period (with lower temperatures) may be due to a compensatory balance between the temperature gradient and evaporation. The incubation time for this nest was 59 days, with 30 chicks hatching simultaneously from the 31 eggs present. These animals were sexed as females by analyzing the external structure of the genitalia at approximately 12 months of age. Subsequently, 24 of these animals had their sex determined through histology of the gonads, confirming the previous results. **Keywords:** *Caiman crocodilus yacare*, nest, incubation temperature, incubation time, sex determination, Pantanal.

Introduction:

Since the importance of temperature in determining sex in reptiles was discovered (Charnier, 1966), many studies have been carried out in an attempt to better understand this mechanism (reviewed by Bull, 1980 and Gutzke, 1987). Most of these experiments have been conducted in the laboratory at constant temperatures throughout the incubation period. Not only can these results not be interpreted for natural nests, as they are subject to weather and temperature fluctuations, but they may also be "hiding" more direct influencing factors (Bull, 1985).

Bull & Vogt (1981) suggest that sex determination depends on a cumulative effect of temperature over the incubation period and that, therefore, the sex ratio is related to the number of hours per day of exposure to a critical temperature and can therefore be related to the combination of average incubation temperature and daily oscillations. Pieau (1982), in an experiment with *Emys orbicularis* turtles, observed large cyclical fluctuations (between 16°C and 40°C) in the temperature of natural nests, indicating the importance of the number of hours of exposure above or below the critical temperature and the influence of the dependence of the rate of embryonic development on temperature.

Little is known about the temperature conditions to which eggs are subjected in their natural nests. Fowler (1979), in an experiment with the common turtle *Chelonia mydas*, found the influence of the number of eggs per nest. In crocodilians, nest incubation temperature patterns vary according to species and nesting sites, but in general the natural sources of nest heating are: sunlight, metabolic heating and decomposition of organic material (Webb & Cooper-Preston, 1989 and Magnuson *et al.*, 1990). Ferguson & Joanen (1982) mapped egg temperatures in 6 natural *Alligator Imississippiensis* nests distributed over 3 different sites: dyke, wetland and dryland. The authors found that nest temperatures varied according to the location and material used to build the nest and also that the position of the egg in relation to the nest determines different thermal conditions, which is important in determining the sex ratio of a spawning. Magnuson (1979) found that environmental thermal fluctuations are responsible for the daily temperature fluctuations in the nests of *Crocodylus porosus*, so, for example, a drop of 3° C in the external temperature would cause a drop of 1° C in the temperature of the nest.

For *Caiman crocodilus yacare*, Campos (1993) found that the incubation temperature of forest nests responds to environmental factors (air temperature, rainfall and time of day) accumulated over the period of 3 to 7 days before hatching and used a model to predict the average temperature of natural nests. In addition, by monitoring the temperature of 3 forest nests for approximately 10 days with measurements at 6-hour intervals, he found a variation of 5°C in these nests. Cintra (1988), working with the same species, through single measurements taken in natural nests, showed that the temperature of the laying chamber would be an average of 33°C.

In this study, the temperature of a natural *C. c. yacare* nest was monitored in an attempt to understand the thermal conditions in which the eggs are subjected during incubation. We sought to answer the following questions:

- What are the daily temperature fluctuations like in the nest?

- How does the internal temperature of the nest respond to environmental variations?
- Does the temperature in the egg chamber differ from the temperature elsewhere in the nest?

Material and Methods:

Area of study:

The Pantanal of Mato Grosso is characterized as a depression in the Upper Paraguay river basin covering an area of 136,700km2 in the states of Mato Grosso and Mato Grosso do Sul, subject to periodic flooding (RADAMBRASIL, 1982a,b and Boggiani & Coimbra, 1996).

The study was carried out in the Pantanal of Mato Grosso do Sul, sub-region of Abobral (Adàmoli, 1982), municipality of Corumbà, in an area of riparian forest of the Miranda River, near the Pantanal Studies Base of the Federal University of Mato Grosso do Sul (BEP/UFMS), located between 19° 34'36"S and 57° 34O6 W."

The nesting season for *C. c. yacare* begins in the rainy season between the end of December and peaks in mid to late January, with mating taking place approximately 70 days before egg-laying (Crawshaw, 1989).

Data collection and analysis:

The experiment began on January 17, 1998. The material used for this experimental study consisted of a recent spawning *of C. c. yacare*, located in a riparian area of the Miranda River. The nest measured 97cm in the long axis, 64cm in the short axis, 41cm high and was 380cm from the water. These dimensions are slightly below the size range described by Crawshaw (1987) who found 120.6 + 19.3cm for the smallest axis and 135.1+22cm for the largest axis and 41.5 + 6.4cm in height. The spawn had 31 eggs, of which 30 chicks hatched simultaneously at 59 days of incubation. When checking the contents of the egg that didn't hatch, no traces of embryonic development were found, indicating that fertilization may not have occurred or that there may have been embryonic death at an early stage.

A week after locating the nest, it was isolated with a wire mesh fence in order to avoid predators. Thermometers were inserted (all accurate to 0.1 °C), three of which were positioned at different depths in the nest and an external thermometer close to the nest, to measure the ambient temperature (Fig. 01) because, according to Magnuson (1979), the air temperature immediately above the surface of the nest is responsible for the gradient of variation within the nest.

Thermometer 1 - egg chamber base (Incotherm 6190, -10 to 50°C)

Thermometer 2 - center of egg chamber (Incotherm 6162, -10 to 50°C)

Thermometer 3 - top of egg chamber (Incotherm 6176, -10 to 50°C)

Thermometer 4 - external temperature (Incotherm 6158, -10 to 50°C)

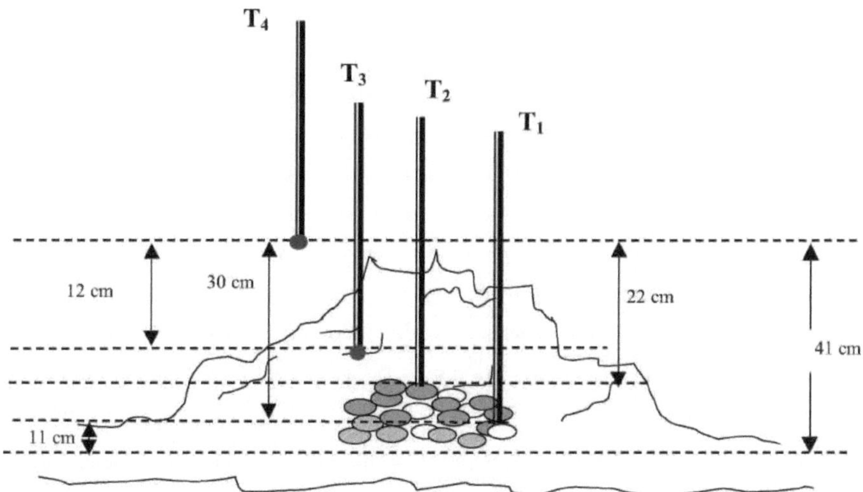

Figure 01 - Schematic of the position of the thermometers in the nest.

For seven consecutive days, 24 hourly temperature measurements were taken. On the following days, temperature measurements were taken between 6am and 6pm. Spawning was monitored for 38 days and at the end of this period the eggs and nest material were transported to the UFMS laboratory in Campo Grande for another experiment. The chicks were sexed by histological examination of the gonads with the collaboration of the Pathological Anatomy laboratory of the Experimental Medicine and Surgery Unit of the Faculty of Medical Sciences of UNICAMP, Campinas, SP.

Immediately after resection, the gonads were immersed in Bouin's solution for four hours, washed quickly in running water and stored in 70gl alcohol in the fridge. Care was taken to keep the identification of the material. In the laboratory, the gonads were sectioned lengthways and placed in Paraplast, keeping the orientation of the cut side. The blocks were processed using 5µ-thick microtome sections, mounted on glass slides and stained with Hematoxylin-Eosin, and then evaluated using light microscopy.

The data was analyzed using ANOVA, Tukey's *post hoc* test, linear regressions and correlations (Statistica software provided by the Brazilian representative in Sao Paulo).

Results:

Analysis of the temperature measurements in the nest reveals the following. The average temperature inside the chamber was 31.89±1.48° C for the entire period. The total average temperatures in the different positions of the nest showed statistically significant differences between each other and with the external temperature ($F3_{.2271=337}.56$ p<0.001).

The Tukey post hoc test applied to the different thermometer reading intervals showed that: the temperature in the center of the laying chamber differed from the ambient temperature and was higher; the average temperature values for most of the readings taken in the center and base of the chamber did not differ statistically significantly from each other. The temperatures at the base and on the surface of the laying chamber did not differ from each other, but were higher than the ambient temperature most of the time. Only in the period between 14:00 and 18:00 did the temperatures of the base and surface of the chamber not differ significantly from the environment. The results of the temperature measurements are shown in Table 01.

Table 1 - Temperature readings taken every hour at the different positions measured (mean ± SD) in the monitored natural nest

Time of day	Position of thermometers in relation to the laying chamber				
	External (T_4)	Surface (T_3)	Center (T_2)	Base (T_1)	ANOVA results $Fgl_{,n};p$
00:00	27,75±1,99	30,31±1,46a	31,97±1,52	31,19±1,12a	$F3_{,28=46},72$ p<0,001
01:00	27,79±0,70	31.43±0.9ac	32.41±0.62bc	31.40±0.51ab	$F3_{,28=55},59$ p<0,001
02:00	27,43±0,79	31,07±0,61a	32,30±0,59	31,29±0,51a	$F3_{,28=79},61$ p<0,001
03:00	26,44±2,67	30,00±3,30b	31,73±1,76a	30.96±1.43ab	$F3_{,32=7},58$ p=0,001
04:00	27,00±0,58	31,14±0,69b	32,34±0,75a	31.44±0.59ab	$F3_{,28=92},21$ gl=3 p<0,001
05:00	27,00±0,58	30,93±0,61b	32,31±0,74a	31.46±0.60ab	$F3_{,28=96},74$ p<0,001
06:00	26,55±2,11	29,32±2,25	31,27±1,79a	30,67±1,34a	$F3_{,148=44}.82$ p<0.001
07:00	26,50±2,02	29,61±2,10	31,58±1,76a	30,95±1,19a	$F3_{,144=56}.42$ p<0.001
08:00	27,06±2,02	29,99±1,65	31,75±1,68a	31,03±1,05a	$F3_{,144=57}.00$ p<0.001
09:00	27,72±2,21	29,86±1,86	31,79±1,64a	30,99±1,18a	$F3_{,144=36}.13$ p<0.001

10:00	28,33±2,11	30,19±1,72b	31,93±1,53a	31.07±1.10ab	$F3_{,144=31}.18$ p<0.001
11:00	29,04±2,16	30,37±1,56b	31,96±1,56a	31.09±1.13ab	$F3_{,140=19}.70$ p<0.001
12:00	29,47±2,21	30,55±1,59b	32,00±1,47a	31.05±1.13ab	$F3_{,144=14}.73$ p<0.001
13:00	30,04±2,14	30,69±1,46b	32,02±1,42a	31.09±1.13ab	$F3_{.144=9}.83$ p<0.001
14:00	30.11±2.61ab	30.71±1.57bd	31,95±1,62c	30.86±1.35ac d	$F3_{.144=6}.14$ p=0.001
15:00	30.14±2.70cd	30.49±1.62ad	32,00±1,53b	30.93±1.19ab c	$F3_{.144=6}.91$ p<0.001
16:00	30.14±2.55ab	30.41±1.53bd	31,86±1,38c	30.83±1.08ac d	$F3_{,143=6},89$ p<0,001
17:00	29,91±2,39a	30.57±1.45ab	31,94±1,29	30,91±1,02b	$F3_{,144=9},85$ p<0,001
18:00	29,55±2,54a	30.26±1.50ac	31,78±1,37b	30.83±1.06bc	$F3_{,144=10}.90$ p<0.001
19:00	29,50±1,52	31.50±0.55bc	32.33±0.50ac	31.30±0.38ab	$F3_{,24=11},37$ p<0,001
20:00	29,33±0,75	31.67±0.52ab	32,32±0,53b	31,33±0,34a	$F3_{,24=32},38$ p<0,001
21:00	29,25±0,76	31.50±0.55ab	32,43±0,47b	31,38±0,62a	$F3_{,24=29},23$ p<0,001
22:00	28,75±0,76	31.33±0.82bc	32.30±0.52ac	31.27±0.45ab	$F3_{,24=32},18$ p<0,001
23:00	28,17±0,75	31.33±0.82ab	32,33±0,54b	31,27±0,45a	$F3_{,24=45},34$ p<0,001
TOTAL	28,65±2,53	30,34±1,69	31,89±1,48	31,00±1,11	$F3_{,2271=337},56$ p<0,001

For each time class, the numbers followed by the same letters do not differ statistically.

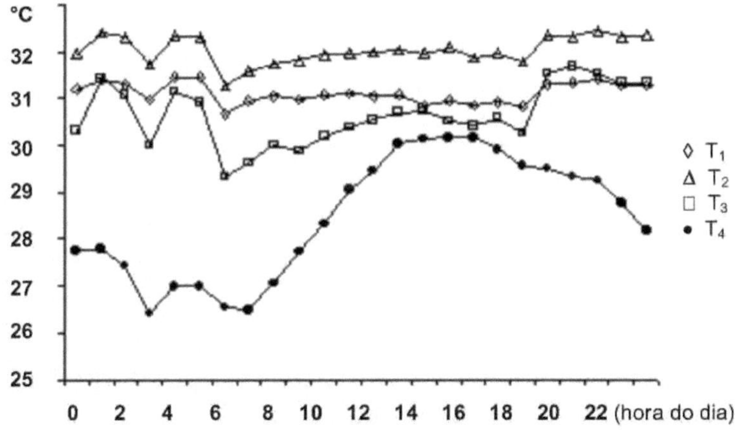

Figure 02 - Average temperatures over 24 hours. Each line in the graph corresponds to the position of the thermometer in relation to the egg chamber. T_1 - temperature of the base of the laying chamber, T2 - temperature of the center of the laying chamber, T3 - temperature of the surface of the laying chamber and T4 external temperature.

Fig. 02 graphically shows the distribution of average hourly temperatures over a 24-hour period. The correlations between the temperature readings obtained from the external thermometer and those from the thermometers placed inside the nest are shown in Fig. 03.

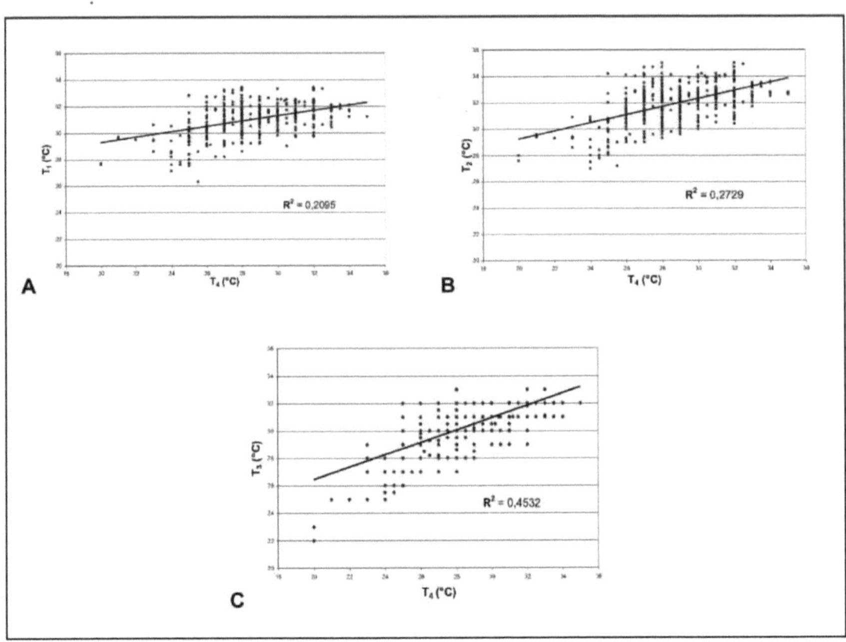

Figure 03 - Correlation of temperatures inside the nest with the outside temperature (T_4). **A** - correlation with the temperature at the base of the laying chamber (Ti), **B** - correlation with the temperature in the center of the laying chamber (T2) and **C** - correlation with the temperature on the surface of the laying chamber (T3).

The difference between the highest and lowest readings obtained from the external thermometer for the whole period was 15° C, while in the center of the chamber it was 8° C, at the base 7.1° C and on the surface 11 oc. The widest range of daily variation was 9°C in the room, 7°C on the surface of the chamber, 5.8°C at the base and 6°C in the center. The variability of the average temperatures for each reading position was: 8.83% in the external temperature; 5.57% on the surface of the chamber; 4.64% in the center and 3.58% at the base.

Of the 30 chicks hatched, 24 became ill and died over the course of 14 months. The chicks recovered from this nest were reintroduced into the Abobral sub-region of the Pantanal in Mato Grosso do Sul. The animals that died were sexed using the histological method described above. The 24 animals evaluated were identified as females.

Discussion:

Analyzing the correlations obtained between the external temperature and the temperatures measured inside the nest (Fig. 03), it can be seen that the internal variations are poorly explained by the external temperature changes (low values of R^2). There is also an indication that the external influence is less the deeper you go into the nest, as the values of R^2 decrease in this order (0.4532, 0.2729 and 0.2092). This observation is indicative of the degree of thermal insulation of the material that makes up the nest, as in the case of the species under study, which uses organic matter (mud, dry leaves and twigs or earth), forming mounds (Crawshaw, 1989 and Marques & Monteiro, 1995). Magnuson (1979) found that the temperature of the environment contributes to the daily variation in the temperature of the nests, but that this variation is small (1.1 + 0.7°C) and is not noticed immediately after an external temperature change.

The variability of the average temperatures inside the laying chamber was lower than outside, indicating greater thermal constancy inside the nest, especially in the lower layer of the chamber. The maintenance and stability of the internal temperature of the nest observed in this study has been observed previously for species with similar nest-building habits. Magnuson & Taylor (1980) drew attention to the importance of the type of material that makes up the nest, finding little thermal variation (between 0.5°C and 0.9°C) in *C. porosus* nests, also built with organic material. For *Crocodylus niloticos* nesting in cavities on sandy beaches, Hutton (1987) monitored the temperature of the nests and found large fluctuations, with drops of up to 10°C in daily temperature and 5°C in average temperature over a 7-day period.

The results of the ANOVA statistical calculations between the measurements of average chamber temperatures, T1, T2 and T3 for the entire period, clearly showed that the different depths of the nest are maintained at different temperatures throughout the incubation period, meaning that for the same clutch the eggs are incubated in different thermal conditions. Ferguson & Joanen (1982), working with *A. mississippiensis* nests, noted the importance of the position of the egg in the laying chamber for the sex determination of the clutch, due to the different thermal conditions that each embryo would be subjected to.

The average total temperature in the center of the chamber remained higher than the external temperature and the temperature measured at other points in the nest (base and surface) throughout the reading period (Fig 02), which indicates the existence of an internal heat source, probably generated through metabolic heating. This result agrees with those reported by Cintra (1988) for *C. c. yacare*, who also found that the temperature in the egg chamber was higher than the external temperature in forest nests. Carr & Hirth (1961), working with *Chelonia mydas*, observed that the metabolic heating of the eggs can raise the temperature of the laying chamber by up to 3°C. Magnuson (1979) found that in the nests *of C. porosus* the temperature remains high and stable inside the laying chamber, through internal heating sources (metabolic heat), even when exposed to a wide temperature gradient between the inside and outside of the nest.

Analyzing the temperatures read at hourly intervals for all positions (Table 01), it can be seen that the period between 14:00 and 18:00 was the only one in which the temperatures taken from the base and surface of the nest did not show statistically significant differences with the ambient temperature. During this period, the sun's incidence must have been greater, causing the external temperature measured just above the nest to increase, bringing it into line with the measurements taken inside.

Fig. 02 also shows that the drop in external temperature at night is not accompanied by a decrease in internal temperature at any point measured in the nest. It would be expected that the temperature gradient, which is higher at night, would cause the interior of the nest to cool down through heat loss to the external environment. It is assumed, however, that this phenomenon is probably offset by a decrease in water evaporation (which is more intense during the day). This balance between the rate of evaporation and conduction should allow the temperature of the nest at night to remain similar to that of the day, as observed in this study.

As for sex determination, all the chicks sexed by histology were females (n = 24), the others had external genitalia similar in shape to the first ones and could also be considered to be of the same sex. The average temperature at which these embryos were exposed varied between 30.34 + 1.69°C and 31.89 + 1.48°C depending on their location in the chamber (Table 01). However, this variation did not seem to have an effect on the sex determination of the chicks.

The results obtained in this study indicate that there is a relatively large amount of thermal insulation, probably due to the shape and type of material used in the nest, allowing a stable internal temperature to be maintained which is higher than that of the environment thanks to the generation of endogenous heat. It is suggested that future studies look into the role of this thermal stability and its influence on the species' sex ratio and embryonic development.

Bibliographical References :

Adâmoli, J. (1982). The Pantanal and its phytogeographic relations with the Cerrados. XXXII National Botanical Congress **Proceedings**:109-119.

Boggiani, P. C. & Coimbra, A. M. (1996). The plains and the wetlands. In P.T.Z. Antas and I. L. S. Nascimento, ed. Tuiuiù - Sob os céus do Pantanal - Biologia e conservaçâo do tuiuiù (Jabiru nycteria). Empresa de Artes, Sao Paulo pp. 18-23.

Bull, J. J. (1980). Sex determination in reptiles. *Quart. Rev. Biol.* **55**:3-21

Bull, J. J. (1985). Sex ratios and nesting temperature in turtles: comparing field and laboratory data. *Ecology* **66**:1115-1122.

Bull, J. J. & Vogt, R. C. (1981). Temperature-sensitive periods of sex determination in Emydid turtles. *J. Exp. Zool.* **218**:435-447.

Carr, A. F. & Hirth, H. F. (1961). Social facilitation in green sea turtle siblings. Animal Behavior 9:68-70.

Charnier, M. (1966). Action de la température sur la sex-ratio chez l'embryon d'*Agama agama* (Agamidae, Lacertillien), Soc. Biol. Quest. Af. **160**:620-622.

Campos, Z. (1993). Effect of habitat on survival of eggs and sex ratios of hatchlings *of Caiman crocodilus yacare* in the Pantanal, Brazil. *J. Herpetology* **27**(2):127-132.

Cintra, R. (1988). Nesting ecology of the Paraguayan Caiman (Caiman yacare) in the Brazilian Pantanal. *J. Herpetology,* **22**(2): 219-222.

Crawshaw, P. G. (1989). Nesting ecology of the Paraguayan (*Caiman yacare*) in the Pantanal of Mato Grosso, Brazil. *MS Thesis (unpubl.).* Gainesville, University of Florida, 68pp.

Ferguson, M. W. J & Joanen, T. (1982). Temperature of egg incubation determines sex in *Alligator mississippiensis. Nature* **296**:850-854.

Fowler, L. E. (1979). Hatchling success and nest predation in the green sea turtle, Chelonia Mydas, at tortuguero, Costa Rica. *Ecology* **60**(5):946-955.

Gutzke, W. H. N. (1987). Sex determination and sexual differentiationin reptiles. *Herpetological Journal* **1**:122-125.

Hutton, J. M. (1987). Incubation temperature, sex ratios and sex determination in a population of Nile crocodiles (*Crocodylus niloticus*). *J. Zool. Lond.* **211:** 143-155.

Magnuson, W. E. (1979). Maintenance of temperature of crocodile nests (Reptilia, Crocodilidae). *J.*

Herpetology **13**(4):439-443.

Magnuson, W. E. & Taylor, J. A. (1980). A description of developmental stages in Crocodylus porosus, for use in aging eggs in the field. Aust. Wildl. Res. 7:479-485.

Magnuson, W. E.; Lima, A. P.; J. M. Hero; T. M. Sanaiotti; E M. Yamakoshi (1990). *Paleosuchus trigonatus* nests: sources of heat and embryo sex ratios. *J. Herpetology* **24**(4):397-400.

Marques, E. J. & Monteiro, E. L. (1995). Ranching of Caiman crocodilus yacare in the Pantanal of Mato Grosso do Sul, Brazil. **In:** Conservation and Management of Caimans and Cocodiles of Latin America. Alejandro Larriera and Luciano M. Verdade. Published by Fundacion Banco Bica, Santo Tomé, Santa Fe, Argentina.

Pieau, C. (1982). Modalities of action of temperature on sexual differentiation in field-developing embryos of the european pond turtle *Emys orbicularis* (Emydidae). *J. Exp. Zool.***220**:353- 360

RADAMBRASIL (1982a). Survey of natural resources. Vol. **27**: Sheet SE. 21 Corumbâ.

RADAMBRASIL (1982b). Survey of natural resources. Vol. **28**: Sheet SF. 21 Campo Grande.

Webb, G. J. W. & Cooper-Preston, H. (1989). Effects of incubation temperature on crocodiles and the evolution of reptilian oviparity. *Amer. Zool.* **29**:953-971.

THERMO-CONTROLLED INCUBATORS FOR EXPERIMENTAL VERIFICATION OF MODELS RELATING EMBRYONIC DEVELOPMENT AND INCUBATION TEMPERATURE.

Rafaela Danielli Nicola[1] , Eliézer José Marques[2] and Jorge H. Nicola[3]

[1] Master's Degree Course in Ecology and Conservation/CCBS, Federal University of Mato Grosso, Brazil.

Grosso do Sul, P.O. Box 649, cEp 79070-900, Campo Grande, Ms.

[2]Department of Biology/ccBs, Federal University of Mato Grosso do Sul.

[3]Solid State Department, Institute of Physics/Unicamp, Campinas, SP.

Summary:

We describe the development and construction of a system composed of independent incubators with computerized control, capable of simultaneously providing different temperature conditions. The system was developed to enable a laboratory experiment to model the influence of temperature variation during the incubation period on determining the sex of *Caiman crocodilus yacare*, and can be used in similar experiments. We built and tested five water bath incubators heated by 2,000 watt electrical resistors. We used thermocouple sensors to read the temperatures. The electrical signal generated by each thermocouple, in the order of a millivolt, is amplified, digitized and multiplexed using specially built hardware that can be interfaced via serial port (Com1,2) with a simple computer, on which appropriately developed software has been installed. The program generates, for each incubator, a temperature reference, with a convenient variation (in this case, sinusoidal, over a period of 24 hours, with amplitudes and average values set independently). Every minute, the program compares the measured temperature with the programmed one, turning on the heaters as necessary. In the laboratory experiment carried out in an air-conditioned room at 22° C, the accuracy verified over a period of more than 60 days was +0.2 C.°

Keywords: incubators, modeling, controlled temperature.

Introduction:

The environment in which a species lives can influence it in different ways and to different degrees. In the case of oviparous reptiles, the first environment experienced by the embryo is the egg, which is affected both by the microenvironment of the nest and by the development of the embryo itself (Congdon *et. al.*, 1995). The importance of temperature in sex determination was first noted in 1966 by Charnier. Since then, the role of incubation temperature in reptile eggs has been the subject of interest by several researchers.

One way of studying sex determination is through laboratory experiments where conditions can be controlled. This work describes the construction of an incubation system that allows flexibility in the choice of temperature range and fluctuation.

The construction of this system was motivated by the knowledge of previous studies that highlight the importance of temperature fluctuations during incubation, acting on sex determination (Bull, 1985; Pieau, 1982 and Bull & Vogt, 1981), and with the more specific aim of testing the model proposed by Georges *et al.,*1994 on a crocodilian species, in this case *Caiman crocodilus yacare*. In this work, the authors incubated eggs of *Caretta caretta* turtles and found that for this species, sex was determined by the proportion of development that occurred at each temperature and not by the total average temperature of the incubation period.

In order to test the model, we needed reliable incubators capable of maintaining a time-varying temperature and which obeyed a sinusoidal function. Nowadays, the low cost of personal computers makes it easy to design incubators with a thermo-control system capable of performing the desired functions.

Material and Method:

Fig. 01 shows one of the five incubators, like water bath tanks, built from galvanized iron sheet. Each tank has a square base measuring approximately 0.80 x 0.80m^2 and a height of 0.30m. In the center of these tanks, insulated electric heaters were installed, of the type normally used for home water heating, 0.26m long and with a power (P) of 2,000 watts at 220 volts. Six aluminum buckets with an average diameter of 28 cm and a capacity of up to 12 alligator eggs or similar, were symmetrically fixed to each tank with the help of aluminum support bars, allowing the water to pass over the entire outer surface of the bucket and constituting the experimental nests themselves. Internally, pipes made of ¾-inch PVC pipe were perforated on the sides of the tanks, making it possible for the water to circulate internally.

Figure 1. Photograph of one of the five tanks that make up the hatchery system built.

The water was collected by a high-flow circulating pump connected to a fixed piping system around the heating element and returned to the tank through peripheral pipes, forming a closed circuit to agitate the water and thermalize the system.

Thermocouples were used to measure the instantaneous temperature of each tank. A schematic representation of how the system works can be seen in Fig. 02.

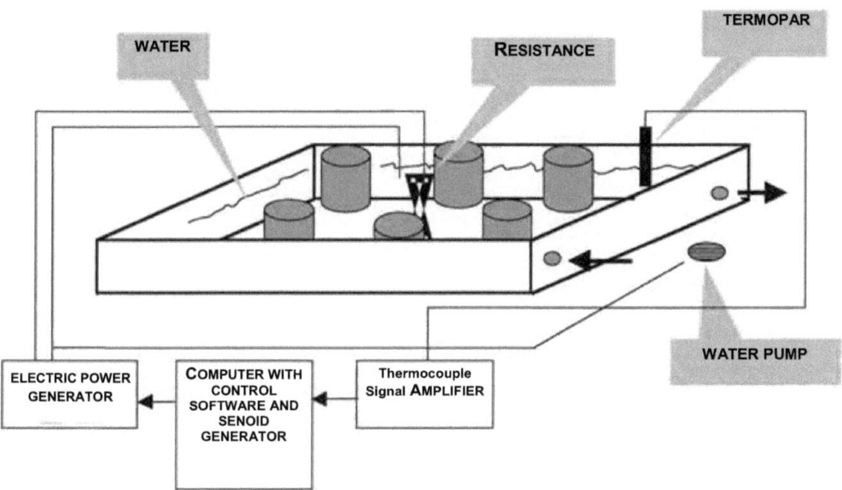

Figure 02 - Schematic of an incubator connected to a computer

The temperature of the system was controlled via the serial port of a 386 computer, interfaced by a specially built analog-digital controller/converter with the thermocouples and electrical resistors.

The temperature was measured using an *Incotherm* mercury thermometer, serial number 58.379, with an accuracy of 0.1^0 C, with a scale of -30^0 C to +50 $C.^0$

Fig. 03 shows the system already installed in a suitable room. As the experiment was carried out during the summer and the Campo Grande region can experience high temperatures during this period, it was necessary to install the incubators in a room that was air-conditioned, keeping the temperature at around 22^0 C, to allow for the desired oscillations.

Figure 03 - Photograph of the incubators installed in an air-conditioned environment.

The feedback system for temperature control was based on theoretical calculations (see table below) for the thermal capacity of each tank and duly adjusted experimentally. Initially, we considered the tanks to be completely full, without the nests and without thermal exchange with the environment. Thus, as the volume (v) of water in each tank is around $0.19m^3$ we have a mass (m) of 190kg. Considering the initial external temperature (t_0) to be 22^0 C, and the temperature to be reached (t_f) to be 30^0 C, the amount of energy required (ΔQ) for this can be put as follows

calculated using the definition of specific heat as detailed below:

$$\Delta Q = m \times (t_f - t_0) \times c_{água},$$

where, in this particular case, $c_{agua} = 1$ cal/g x^0 C is the specific heat of water for the temperature region of interest. Using the equivalence factor between calories and joules, we **obtain**:

ΔQ = 6,362,720 joules,

and, considering that:

$\Delta Q = P \times T$,

where T is the time taken to reach the temperature tf, and P is the resistance used, we know that P= 2,000 watts:

T = ΔQ /P = 3,181s approximately, or 53 minutes, equivalent to approximately 40s for increases of 0.1^0 C, corresponding to the precision of the system as a whole.

The temperature control can follow any previously programmed time function. In this case, a sinusoidal function was chosen with a 24-hour period and adjustable average and maximum values. The start of the experiment can also be adjusted by the program. The specially developed software generates five sinusoids as a function of time, which is measured internally using the computer's clock. These curves are normalized in degrees Celsius x seconds and previously calibrated using the precision thermometer, forming the values of the expected instantaneous temperatures (t_1, t_2, t_3, t4 and t_5) in each of the five incubators. The electrical signals generated by the thermocouples are read every minute by the computer and compared with the values t_1, t_2, t_3, t4 and t5 respectively. From this comparison, depending on whether the temperatures read are above or below the respective expected values, the program decides to maintain the electrical resistances (individually) turned off or on, in order to maintain the correct temperature in each incubator.

Results and Discussion:

The system developed is heated by electrical resistors, while cooling occurs spontaneously due to the difference in temperature between the incubators and the environment. The theoretical calculations presented above served to indicate the practicality of the project, and adjustments were made according to experimental measurements, guaranteeing and adapting the functioning of the thermal control program.

For the system assembled, as shown in Fig. 2, and in the range 24 - 350C, the average time measured to cool 0.1^0 C was 30s, while the average time to raise 0.1^0 C was 52s. Comparing this last result with the theoretical value of 40s shows a difference of 12s. This difference is to be expected since the theoretical calculations did not take into account factors such as the existence of nests (buckets), thermal dissipation from the sides and bottom of the tanks and thermal loss through surface water evaporation.

The measured values of 30s to cool down and 52s to heat up 0.1^0 C represent the thermal inertia of the system as a whole. As the control program reads every 60s, it can happen that the temperature decreases by up to 0.2^0 C below the nominal value or increases by at most a little more than 0.1^0 C above the correct value, guaranteeing a maximum error of 0.2 C.^{0}

The incubators, numbered from 1 to 5, were tested for a period of 60 days, with no.° 1 being kept at a constant temperature of 30^0 c, no.° 2 oscillating periodically between 29 and 31^0 C, no.° 3 oscillating between 28 and 32^0 C, no.° 4 between 27 and 33°C and no. 5 between 26 and 34^0 C. During this period, at random times, we took periodic measurements of the temperature of the environment and the water, and found that the discrepancy with the expected value did not exceed 0.2 C.^{0}

Aluminum, which is a good conductor of heat, was used in the buckets used for the experimental nests, allowing for the rapid thermal equilibrium of the environment of the eggs, which in turn were deliberately placed in contact with the inner wall in order to facilitate this process.

Conclusions :

Computer-controlled temperature incubators, built to test egg incubation models, are practical, economical to build and reliable. The capacity of approximately 60 eggs per tank allows for a statistically reliable evaluation of biological data dependent on incubation temperature. With minor modifications to the controller program, new tanks can be added, increasing the number of incubation situations and expanding the system's capacity.

Bibliographical References :

Charnier, M. (1966). Action de la température sur la sex-ratio chez l'embryon d'*Agama agama* (Agamidae, Lacertillien), *Soc. Biol. Quest. Af.* **160**:620-622.

Bull, J. J. (1985). Sex ratios and nesting temperature in turtles: comparing field and laboratory data. *Ecology* **66**:1115-1122.

Bull, J. J. & Vogt, R. C. (1981). Temperature-sensitive periods of sex determination in Emydid turtles. *J. Exp. Zool.* **218**:435-447.

Congdon, J. D.; Fischer, R. U. & Gatten, R. E. Jr. (1995). Effects of incubation temperatures on characteristics of hatchling American alligators. *Herpetologica* **51** (4) 497-504.

Georges, A.; Limpus, C. & Stoutjesdijk, R. (1994). Hatchling sex in the marine turtle *Caretta caretta* is determined by proportion of development at a temperature, not daily duration of exposure. *J. Exp. Zool*. **270**:432-444.

Pieau, C. (1982). Modalities of action of temperature on sexual differentiation in field-developing embryos of the european pond turtle *Emys orbicularis* (Emydidae). *J. Exp. Zool*. **220**:353360

EFFECT OF INCUBATION TEMPERATURE, UNDER CONTROLLED CONDITIONS, ON THE SEX RATIO OF THE MARSH ALLIGATOR, *CAIMAN CROCODILUS YACARE.*

Rafaela Danielli Nicola[1] , Eliézer José Marques[2]

1 Master's Degree Course in Ecology and Conservation/CCBS, Federal University of Mato Grosso do Sul, P.O. Box 649, CEP 79070-900, Campo Grande, MS. 2 [2]Department of Biology/CCBS, Federal University of Mato Grosso do Sul.

Summary:

In this study, we incubated marsh caiman (*Caiman crocodilus yacare*) eggs in five different temperature-controlled treatments, with equal averages (30°C) and different fluctuations. The aim of this study was to test a model of daily temperature fluctuation, suggested for species that show a pattern of temperature dependence in sex determination (TSD). For incubation in the laboratory, 72 eggs of *C. c. yacare* (within five days of laying) from 14 spawnings were used for each treatment. A total of 253 chicks hatched, and sex determination was carried out through histology of the gonads (N=180) and by comparing the external structure of the genitalia. All the chicks were identified as females. For the fluctuation ranges tested, the average incubation temperature was a good indicator of sex ratio, and no influence of temperature fluctuations on sex determination was found.

Keywords: *Caiman crocodilus yacare*, daily temperature fluctuation, TSD, sex determination, incubation temperature.

Introduction:

For many species whose sex definition takes place after conception and is influenced by the environment (ESD), incubation temperature stands out from other factors, playing a major role. According to Bull (1983), different patterns of temperature dependence can be found in sex determination. In lizards and alligators, higher temperatures produce males while lower temperatures produce females, but in several species of turtles the pattern is the opposite. Another pattern found is when females occur at low and high temperatures and males at intermediate temperatures, the latter can be seen in crocodiles. Lang & Andrews (1994) and Deeming & Ferguson (1989) propose that this is the primitive pattern from which the others derive.

The sex ratio of temperature-determined species (TSD):

Temperature sex dependence), is biased towards one sex in most of the incubation temperatures studied, i.e. most incubation temperatures result in individuals of only one sex. In a narrow temperature range, male and female individuals are generated in different proportions (Bull, 1983). The temperature range that results in an equal number of male and female individuals is called pivotal (Yntema & Mrosovsky, 1980 and Mrosovsky, 1980).

No heteromorphism of sex chromosomes has been found in crocodilians (Cohen & Gans, 1970), which seems to indicate that the environment is responsible for sex determination in all species of the group. Ferguson (1985) attributes the temperature at which the eggs are exposed during incubation as a universal factor in the sexual definition of crocodilians. Deeming & Ferguson (1989), proposed a hypothesis for the sex determination of crocodilians, according to which, by default, embryos develop into females unless they are exposed to ideal thermal conditions for the formation of males (male determination factor). The time and intensity of this exposure required to alter the sex ratio would vary according to the species and the intrinsic characteristics of each individual. Bull & Charnov (1989) present a hypothesis to explain the evolution of environmental sex determination (ESD) and sex ratio bias. In their work, the authors postulate that ESD will be favored whenever the environment has a different influence on the fitness (measured by fertility) of the sexes.

Temperature also seems to act on other biological and behavioral functions of crocodiles after birth, such as thermoregulation (Lang 1987), growth and survival (Ferguson & Joanen, 1982; Joanen *et al.*, 1987 and Hutton, 1987), thermal choice of environment and pigmentation (Deemig & Ferguson, 1989).

As for embryonic development, there is a temperature range in which development occurs under normal conditions, while incubation at temperatures at the extremes of this range can result in anomalies (Whitehead *et al.*, 1990). Studies on embryonic development use different measures to evaluate it. Thus, embryonic development can be assessed in different ways, such as changes in body mass, the formation of tissues and organs, incubation time or aspects of shape.

For both crocodilians and other reptiles with TSD, several studies indicate that the rate of embryonic development increases, within a certain limit, with temperature (examples: Bustard & Greenham, 1968; Ferguson, 1987; Webb *et al.*, 1987a; Deemig & Ferguson, 1989). The rate of embryonic development, expressed as a variation in daily biomass, generally shows exponential dependence on temperature, also within certain limits. By using different parameters to measure development, some authors have obtained another type of behavior of the embryonic development rate with temperature.

Whitehead *et al.* (1990) found that, for *C. johnstoni*, embryonic development throughout the incubation period, at a known constant temperature, is most intense between 70% and 80% of the period. Also in this work, comparing incubations at different temperatures, the authors found that the growth constant increases linearly with temperature and, by comparing the inverse of the total incubation times in different thermal conditions, considering them as the rate of embryonic development, they also found an increasing linear relationship.

Ferguson (1987) tried to classify the embryonic development of two crocodilian species (*C. johnstoni* and *C. porosus*) into stages, according to morphological rather than chronological characteristics (incubation time), thus allowing generalizations to other species. In his work, the author found that the rate of development at different temperatures varies according to the tissue analyzed and that higher temperatures cause an acceleration in development during the initial stages of organogenesis (up to stage 20). In the later stages, this relationship becomes less evident.

Variations in linear measurements of the embryo with incubation time (e.g. length x time) can also give an indication of embryonic development over time, showing a very significant linear relationship (Georges *et al.*, 1994).

The way in which temperature influences the sex definition of the embryo has not yet been clarified. Bull and Vogt (1981) suggest that sex determination depends on a cumulative effect of temperature over the incubation period and that, therefore, the sex ratio is related to the number of hours per day of exposure to a critical temperature, and may therefore be related to a combination of average incubation temperature and daily oscillations. Pieau (1982), in an experiment with *Emys orbicularis* turtles, observed large cyclical fluctuations between 16°C and 40°C in incubation temperature; proposing that the number of hours of exposure of the eggs to temperatures above or below the critical temperature, associated with the form of dependence (also with temperature) of the embryonic development rate, is the determining factor in defining the sex of the embryo.

In Bull's (1985) work, he suggests that sex determination depends directly on the amount of embryonic development carried out in a given thermal condition, so that temperature acts indirectly on sex. Also in this work, the author points to the possibility of developing a model to verify this hypothesis.

Along these lines, Georges *et al.* (1994) presented a mathematical model that made it possible to experimentally verify the hypotheses of Bull (1985), Pieau (1982) and Bull & Vogt (1981) in a rigorous manner. In this work, Georges *et al.* (1994) proved, through an incubation experiment at varying temperatures according to a sinusoidal pattern with *Caretta caretta* turtles, that sexual differentiation depends on the daily proportion of embryonic development that occurs at each temperature, and not directly on the average incubation temperature throughout the period. They also showed that it is possible to replace the oscillating temperature system with a constant temperature equivalent (CTE), calculated using the model.

The successful application of the model in the experiment by Georges *et al.* (1994) for turtles suggests the possibility of extending it to understand the action of temperature on the nests of other species which are not normally subject to daily temperature oscillations. In this case, by forcing incubation in the laboratory with a standard oscillating temperature (sinusoidal, for example), conditions are created for analyzing Bull's (1985) hypotheses.

Experiments carried out on natural nests of *C. c. yacare* indicate that they are subject to small daily thermal oscillations, keeping the temperature inside the laying chamber stable (Campos, 1993; Cintra, 1988 and Chapter I of this work).

Several studies indicate that the embryonic development of crocodilians increases with increasing temperature (Deemig & Ferguson, 1989, Ferguson, 1987; Webb *et al.*, 1987), and the same should be true for the marsh caiman. As has been pointed out, under natural conditions, the eggs of *C. c. yacare* are not subject to great thermal differences, but the dependence of temperature on sex determination and the speed of embryonic development motivates us to carry out experiments in an attempt to better understand the functioning of the TSD mechanism for the species. Our proposal is to test the hypothesis raised by Bull (1985) and confirmed by Georges *et al.* (1994) for *C. caretta*, using the same model, in order to verify its applicability to crocodilians.

In the present study, through an experiment on the incubation *of C. c. yacare* eggs at variable and controlled temperatures, we tested, using the model proposed by Georges *et al.* (1994), the hypothesis that the proportion of embryonic development occurring at each temperature is the determining factor of sex in embryos of this species. In order to establish the initial conditions for the study, we used data from the literature on constant temperature incubation of the Pantanal caiman.

Material and methods

Area of Study

Collections were made in the Pantanal region of the Miranda and Abobral sub-regions in the municipality of Corumbà, close to the research base (BEP) of the Federal University of Mato Grosso do Sul, in areas of the capes and ridges of the Miranda and Vermelho rivers (Adàmoli, 1982).

We visited the area by boat and on foot during the species' nesting season, at the beginning of January, as indicated by studies by Cintra (1988), Crawshaw (1987) and Campos (1993). We located nests that had not yet been laid or had recently been laid. When a nest was found, an egg was removed to estimate the incubation time using the width of the opaque band that forms on the eggshell due to structural changes and water loss caused by the dissolution of calcium carbonate crystals in the shell (Webb *et al*, 1987b). Fourteen spawns were collected along with nest material, with incubation times ranging from one to five days. The original position of the eggs was indicated with a pen mark and maintained during handling to prevent the embryo from detaching or rotating, which could lead to its death (Webb *et al.*, 1987b).

Incubation

The eggs were transported to the laboratory at UFMS in Campo Grande where they were measured, weighed and individually identified using a numbered plastic plate attached to gauze. These eggs were sorted by ovoscopy (Pinheiro *et al.*, 1997), making it possible to discard those that did not show the formation of an opaque band. Broken eggs were also discarded.

The eggs from each *C. c. yacare* spawning, together with the nest material, were arranged proportionally (approximately 5 eggs from each spawning per incubator) in the five different temperature treatments, in five-liter aluminum buckets with a plastic mesh structure at the bottom, allowing excess water to be deposited without soaking the eggs. To avoid contamination by fungi or bacteria, the organic matter used as a substrate for incubation was sterilized in an oven at 80°C for 12 hours.

Five water bath incubators were used, as described in Chapter II, measuring 0.80m x 0.80m and 0.30 meters high. The incubators were installed in an air-conditioned room with a constant outside temperature of 22°C. Each incubator had six aluminium buckets immersed in a water bath, with the capacity to house 12 eggs so that they all had the greatest surface area in contact with the bucket wall, allowing equal thermal conditions per treatment.

Handling puppies:

After birth, the chicks were handled according to the methodology developed by Marques & Monteiro (1995). The chicks were immediately washed under running water and treated with Rifocin Spray (Rifamicin SV, Merrell Lepetit Farmacèutica LTDA) to prevent infections. For identification purposes, they were numbered with a color code attached to the second simple crest of the tail.

The animals were then weighed and measured (total length, body length, skull length and weight). The morphometric results will be presented in the next chapter.

Sexual Identification:

The sexual identification of the cubs was first done through visual inspection of the genitalia at approximately six months of age. The animals were being kept in captivity to study their development, but towards the end of this period (10 months) they began to show symptoms of an unknown pathology, and despite our best efforts, many of them died (n = 170). Biological materials from these animals were preserved for later identification of the pathology and the gonads were used for histological sexing. The gonad tissues were fixed in Bouin for 4 hours, transferred to 70gl alcohol and cooled. After fixation, the material was sectioned and analyzed by light microscopy. The slides were prepared at the Pathological Anatomy Laboratory of the Experimental Medicine and Surgery Unit of FCM/UNICAMP, Campinas, São Paulo.

For the results of this study, only animals that were sexed by histology were considered. The animals that survived were released in the Pantanal, Miranda and Abobral sub-regions, after visual inspection of the genitalia.

Fig. 01 shows photographs of a female gonad from an animal incubated under laboratory conditions.

Figure 01 a - Female gonad seen through a surgical microscope (x6), arrow showing the structure of the duct.

Figure 01 b - Histological aspect of the female gonad, arrow showing oocytes.

Figure 01 c - Histological aspect of female gonad, arrow A showing duct and arrow B indicating oocytes.

Theoretical Background

The mathematical model developed by Georges *et al.* (1994), to prove the effect of the temperature dependence of the embryonic development rate on the definition of the sexes of hatchlings, is very simplified and should be viewed with caution when applied to a real biological system. In the specific case of *C. caretta* turtles, the model and experimental results showed very good agreement.

The equation of this model considers that the embryonic development rate **(ds/dt)** of the species under study as a function of temperature (T) shows linear behavior, at least in the incubation temperature region, and is represented by the equation of a straight line as follows:

$$ds/dt = b(T - T_0) \quad \textbf{equation 01}$$

where:

b parameter defining linearity

T_0 a critical temperature characteristic of the system under study,

shown in the graph below as the temperature value for which the embryonic development rate is zero.

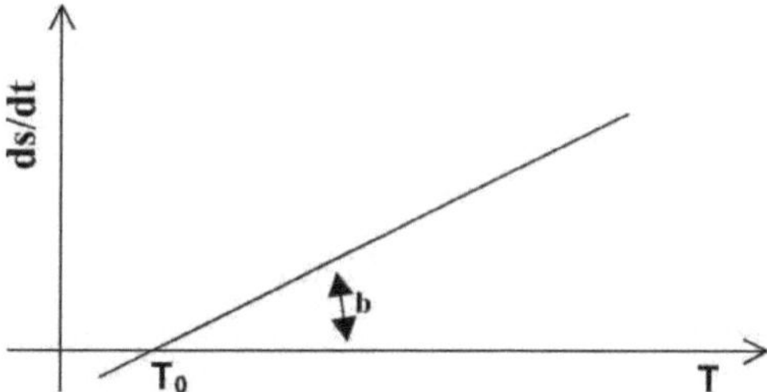

Figure 02 - Graphical representation of equation 01

The design of a laboratory experiment makes it possible to set variations in temperature (T) during the incubation period according to the objectives of the experiment.

One possibility would be to maintain the temperature at a constant value, in which case, according to the figure above, the rate of embryonic development would be the same during incubation. In this case, the experiment would not contribute to understanding the effect of the temperature dependence of the rate of embryonic development on defining the sexes of the offspring.

It can also be assumed that the incubation temperature (T) varies periodically with time (t), according to a 24-hour sinusoid, represented by equation 02 and Fig. 03.

$$T = R.\cos(t) + M \quad \textbf{equation 02}$$

Where:

R is the amplitude of the temperature oscillation

M is the average temperature

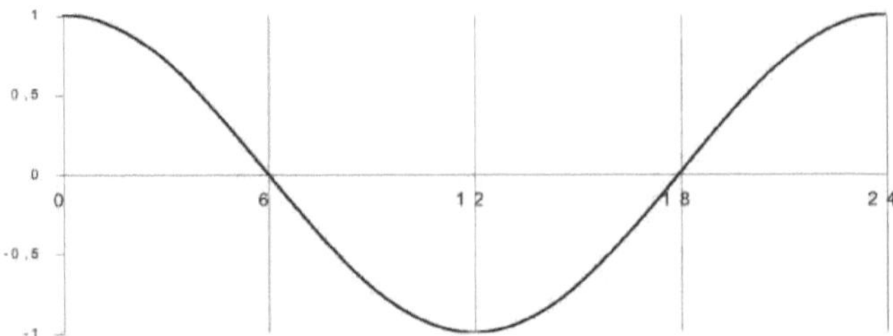

Figure 03 - Temperature over 24 hours

The use of this temperature variation as an experimental model is justified since the cosine (or sine) function makes it possible to use simple mathematical treatments, allowing numerical results for comparison between experiments and also facilitating the interpretation of the results obtained (Georges *et al.*, 1994).

For this simplified model it is easy to see that at certain times of the period, when the temperature is above average, the rate of embryonic development is higher, contributing more to the final result of the incubation process than at times of the period when the temperature is below average (M). So the average temperature only has mathematical significance and an experiment carried out at a constant temperature Tc = M will not be equivalent to one carried out under oscillating temperature conditions, represented by a sine wave with the same average Tc = M.

Georges *et al.* (1994) showed that, for each incubation process with variable temperature according to equation 02, when ds/dt is linear with temperature (eq. 01), a characteristic temperature for the process can be calculated (CTE, explained above), which defines the temperature to be used in an experiment at constant temperature, aiming for results equivalent to those obtained under conditions of variable temperature according to the sinusoidal.

The calculations, detailed in Georges *et al.* (1994), show that the CTE can be obtained from the values of R, M, and T_0 , (defined above). Although this calculation requires the application of an interactive mathematical process, obtaining the CTE is relatively easy.

Monitoring and incubation temperatures:

Spawning was monitored throughout the incubation period, adding organic material from natural nests whenever necessary.

Humidity was maintained daily by sprinkling water directly into the buckets (artificial nests). The thermal conditions of each incubator were previously established, as shown in table 01.

Table 01 - Thermal conditions of each incubator

Incubator	Temperature
01	Constant 30°C
02	30°C +1°C
03	30°C +2°C
04	30°C +3°C
05	30°C +4

Incubator number (treatment) and oscillation amplitude per treatment.

The choice of temperature conditions for the five different incubators was made with the expectation that at least for incubators 04 and 05 the CTE would reach values within the transition temperature range, i.e. between 31°C and 34°C (table 02).

Table 02- Expected CTE values for T0 =26°C.

Incubator	R (°C)	M (°C)	Expected ETC
01	0	30	30
02	1	30	30,24
03	2	30	30,88
04	3	30	31,72
05	4	30	32,68

R - oscillation amplitude, M - average temperature, T_0 - critical temperature for embryonic development, CTE - equivalent constant temperature for these conditions.

We also assumed, for the calculations in Tab. 02, that the temperature below which embryonic development does not occur would be 26°C. This arbitrary choice, based on experiments with other species, was necessary as a starting point for the present experiment.

In a laboratory experiment with *Caiman crocodilus yacare*, Pinheiro *et al.* (1997) found a pattern of response to temperature with a unimodal distribution in which eggs incubated at low temperatures (below 31°C) resulted only in females, while eggs incubated at high temperatures (above 34°C) resulted only in males. The pivotal temperature would be 33°C and the transition range, which would cause males and females to hatch in different proportions, would be between 31°C and 33°C. Thus, according to table 02, taking into account the temperature values found in the work by Pinheiro *et al.* (1997), incubators 04 and 05, with the greatest fluctuations, would result in CTE values within the transition range, which should allow the development of males and females.

Results:

Table 3 summarizes the results of interest for this study. For the five treatments, all the gonads analyzed by histology showed female characteristics with the presence of oocytes (Fig. 01); thus the sex ratio was the same between hatcheries (100% female). The hatching rate was lowest for incubator 01 (48%) while incubators 03 and 05 showed the highest rates (85%). For all treatments, the number of chicks sexed by histology exceeded half the number hatched.

Of the hatched chicks, those that were not sexed by histology were evaluated by direct inspection of the external genitalia (at approximately 6 and 12 months of age) and showed no differences in relation to the external genitalia of the individuals that had female gonads. Apparently, even the individuals that were not counted in the sexing in Table 03 were also female.

Table 03 - Total number of eggs incubated, average incubation time, percentage of chicks hatched, percentage of chicks histologically sexed and sex ratio.

Incubator	**Hatched eggs** (number)	**You can incubation** (mean + SD in days)	**Puppies** (Total N and % hatching)	**Sexed** (Total N and % of sexed chicks)	**Sex** (female: male)
01	72	77,7 + 4,72	35 (48%)	27 (77%)	1:0
02	72	77,0 + 5,07	44 (61%)	33 (75%)	1:0
03	72	78,1 + 4,86	61 (85%)	40 (66%)	1:0
04	72	81,1 + 4,97	52 (72%)	32 (62%)	1:0
05	72	78,7 + 5,11	61 (85%)	38 (63%)	1:0
Total	360	79,6 + 5,48	253 (70%)	170 (67%)	1:0

The average incubation time for all chicks was 79.56 + 5.48 days, with the first hatch occurring at 65 days and the last at 92 days. The comparison of incubation periods between treatments showed that there were differences only between the incubation periods of incubator 04 and incubators 01, 02 and 03 (ANOVA and Tukey's post hoc test with $F_{4, 245} = 4.36$ $p < 0.05$). Several chicks born from the 83rd day of incubation onwards showed malformation problems (they were born with the calf exposed, probably caused by omphalitis, as described by Matushima & Ramos (1995)) and did not show normal embryonic development. Thus, recalculating the average incubation time for chicks born between 65 and 82 days gives 77.00 + 4.45 days. In the specific case of the constant temperature incubator at 30 °C, this calculation leads to a corrected average incubation time of 77.13 + 4.58.

Discussion:

With the data available in the literature at the start of this work, it was possible to construct Table 02 for the design of this laboratory experiment. If the model were applicable to the marsh caiman, we would have predicted different sex ratios for the five proposed treatments. Thus, for incubator 05, with a CTE of 32.68°C, the sex ratio would be approximately 1:1, i.e. equal numbers of males and females. In incubator 04, a CTE of 31.72°C would result in more than 13% males, incubator 03 should have less than 13% males and incubators 02 and 01 would result in only female chicks (considering the experiment by Pinheiro *et al.*,

1997).

The experimental results, as shown in Table 03, differ from what was expected as the sex ratio obtained was 1:0 (females/males) for all treatments, so a careful analysis is needed to interpret these results in the light of the theory under discussion.

The factors that may have contributed to this difference between the results obtained and the results expected can be classified as follows:

J Due to data acquisition errors.

J Due to the non-linearity of ds/dt.

J Due to the inappropriate choice of initial parameters for the CTE calculation (To = 26°C and transition range between males and females of 31°C ≤ T ≤ 33°C)

Errors in data acquisition can be ruled out since the experiment was carried out with the help of correctly calibrated equipment, with a sensitivity higher than that required by the experiment and monitored constantly. Sexing was carried out through histology of the gonads (Fig. 01), allowing the visualization of oocytes, leaving no doubt as to the results obtained.

The effect of a non-linearity for ds/dt in the temperature range covered by the experiment (M + R) may also have interfered with the initial CTE calculation. This factor is predicted by the extension of the George *et al.* (1994) model, assuming non-linear ds/dt, however, for quantitative evaluation it is necessary to know ds/dt as a function of T for the entire temperature interval. This information is not found in the literature.

The choice of parameters used to calculate CTE (To = 26°C and 31°C ≤ T ≤ 33°C, sex ratio transition range) may have been responsible for the discrepancy between the results expected and those obtained. Although the study by Pinheiro *et al.* (1997) found the transition temperature range for *C. c. yacare* through an appropriate laboratory experiment, the egg sample collected for this study came from another sub-region of the Pantanal and from another nesting season. We cannot guarantee that the temperature dependence relationships are identical between the two populations.

The To value used to develop the model with the Pantanal caiman may also differ between species and between different populations. Considering specific data on the incubation time of *C. c. yacare* obtained in this work and from monitoring the temperature of a nest under natural conditions, we have: 59 days of incubation at a temperature of 31.89 + 1.48°C (chapter I) and 77.7 + 4.72 days for an incubation temperature of 30°C. Using this information and the methodology used by Whitehead *et al.* (1990) to calculate the embryonic development rate as a function of temperature in *C. johnstoni*, we can calculate the embryonic development rate (ds/dt in days^{-1}) for the Pantanal caiman, as shown in Table 04.

Table 04 - embryonic development rate for two temperatures.

T (°C)	**Incubation days**	**ds/dt** (days^{-1} x 10)$^{-3}$
30	77,13	12,9
31,89	59	16,9

These results for the embryonic development rate *of Caiman crocodilus yacare*, at two different temperatures, allow a first approximation for the calculation of To, specific to the species under study. This calculation is made by extrapolating the linear function that represents the embryonic development rate, leading to the calculated value of To = 23.85 °C

This new value for To, available only after completing the measurements of interest for this TSD study, now allows a new CTE calculation for the temperature oscillation conditions defined for the experiment (M, R). These results are shown in Table 05.

Table 05 - CTE values calculated for *C. c. yacare,* considering To = 23.85 °C.

Incubator	**R** (°C)	**M** (°C)	**Calculated ETC** (°C)
01	0	30	30
02	1	30	30,16
03	2	30	30,62
04	3	30	31,27
05	4	30	32,12

Although the To value has been corrected using the results obtained here for *C. c. yacare*, it can be seen that the recalculated CTE values change little, indicating that variations in To have little influence on the model. Thus, any inaccuracy in the calculation of To, due to the lack of information on the rate of embryonic development with temperature, must not have interfered significantly with the CTE values.

The results of this work, females at low average temperature, are in line with information on the sexual differentiation of alligators, which follow a unimodal pattern with females also at low temperatures Bull (1983).

Therefore, based on the above considerations, the experimental results of the present work under conditions of fluctuating incubation temperatures lead to two independent but not mutually exclusive hypotheses:

J The rate of embryonic development for the species shows little temperature-dependent variation within the range of thermal oscillation studied.

J The mechanisms that determine the sex of the marsh caiman are more directly related to temperature and not to the rate of embryonic development. In this case, the determination of males may be related to exposure time and temperature intensity. In the present experiment, exposure to oscillating temperatures, with sinusoidal fluctuations, may not have been sufficient to provide conditions for the embryos to develop into males. Deeming & Ferguson (1989), in a hypothesis proposed for the sex determination of crocodiles, suggest that the birth of females is always favored, unless there is a "male determination factor" acting on the embryo.

Biographical references:

Adàmoli, J. (1982). The Pantanal and its phytogeographic relations with the Cerrados. XXXII National Botanical Congress **Proceedings**:109-119.

Bull, J. J. (1983). Evolution of sex determining mechanisms. Chapter 9. Edited by The Benjamin/Cummings pbsh. Co. Menlo Park, California.

Bull, J. J. (1985). Sex ratios and nesting temperature in turtles: comparing field and laboratory data. *Ecology* **66**:1115-1122.

Bull, J. J. & Charnov, E. L. (1989). Enigmatic reptilian sex ratios. *Evolution*, **43**(7):1561-1566.

Bull, J. J. & Vogt, R. C. (1981) Temperature-sensitive periods of sex determination in Emydid turtles. *J. Exp. Zool.* **218**:435-447.

Campos, Z. (1993). Effect of habitat on survival of eggs and sex ratios of hatchlings *of Caiman crocodilus yacare* in the Pantanal, Brazil. *J. Herpetology* **27**(2):127-132.

Cintra, R. (1988). Nesting ecology of the Paraguayan Caiman (Caiman yacare) in the Brazilian Pantanal. *J. Herpetology*, **22**(2): 219-222.

Cohen, M. M. & Gans, C. (1970). The chromosomes of the order Crocodilia. *Cytogenetics* **9**:81105.

Crawshaw, P.G. (1987). Nesting ecology of the Paraguayan (*Caiman yacare*) in the Pantanal of Mato Grosso, Brazil. MS Thesis (unpubl.). Gainesville, University of Florida, 68pp.

Deeming, D. C. & Ferguson, M. W. J. (1989). The mechanism of temperature dependent Sex determination in Crocodilians: a hypothesis. *American Zoology* **29**:973-985.

Ferguson, M. W. J. (1985). The reproductive biology and embryology of crocodilians. **In:** Biology of Reptilia. 14 Development A:329-49. Gans G. Billett, F. & Maderson P. F. A. (Eds) New York: John Wiley & Sons.

Ferguson, M. W. J. (1987). Post-laying stages of embryonic development in crocodilians. Chap.

45:427-444 **in**: Wildlife management: Crocodiles and Alligators G. W. Webb; C. Manolis and P. J. Whitehead, pblsh. Surrey Beatty & sons Pty Limited.

Ferguson, M. W. J. & Joanen, T. (1982). Temperature of egg incubation determines sex in *Alligator mississippiensis*. Nature, **296:**850-854.

Georges, A. Limpus, C. & Stoutjesdijk, R. (1994). Hatchling sex in the marine turtle *Caretta caretta* is determined by proportion of development at a temperature, not daily duration of exposure. *J. Exp. Zool.* **270**:432-444.

Hutton, J. M. (1987). Incubation temperature, sex ratios and sex determination in a population of Nile crocodiles (*Crocodylus niloticus*). *J. Zool. Lond.* **211:** 143-155.

Joanen, T.; McNease, L. & Ferguson, M. W. J. (1987). The effect of egg incubation temperature in post hatchling growth of American Alligators. **In:** Wildlife management: Crocodiles and Alligators G. W. Webb; C. Manolis and P. J. Whitehead, pblsh. Surrey Beatty & sons Pty Limited.

Lang, J. W. & Andrews, H.V. (1994). Temperature-dependent Sex determination in Crocodilians. *J. Exp. Zool.* **270:** 28-44.

Lang, J. W. (1987) Crocodilian thermal selection. . **in**: Wildlife management: Crocodiles and Alligators G. W. Webb; C. Manolis and P. J. Whitehead, pblsh. Surrey Beatty & sons Pty Limited.

Marques, E. J. & Monteiro, E. L. (1995). Ranching of Caiman crocodilus yacare in the Pantanal of Mato Grosso do Sul, Brazil. **In:** Conservation and Management of Caimans and Cocodiles of Latin America. Alejandro Larriera and Luciano M. Verdade. Published by Fundacion Banco Bica, Santo Tomé, Santa Fe, Argentina.

Matushima, E. R. & Ramos, M. C. C. (1995). Some pathologies in alligator breeding in Brazil. **In:** La conservacion y el manejo de caimanes y cocodrilos de America Latina. Alejandro Larriera and Luciano M. Verdade. Published by Fundacion Banco Bica, Santo Tomé, Santa Fe, Argentina.

Mrsovsky, N. (1980) Thermal biology of sea turtle. *Amer. Zool.* **20**:531-547

Pieau, C. (1982). Modalities of action of temperature on sexual differentiation in field-developing embryos of the european pond turtle *Emys orbicularis* (Emydidae). *J. Exp. Zool.* **220**:353360

Pinheiro, M.; Mourao, G.; Campos Z. & Coutinho M. (1997). Influence of incubation temperature on sex determination in alligators (*Caiman crocodilus yacare*). *Rev. Brasil. Biol.* **57**(3):383-391.

Webb, G. J. W; Beal, A. M.; Manolis, S. C. & Dempsey, K. E. (1987a). The effect of incubation temperature on sex determination and embryonic development rate in *Crocodylus johnstoni* and *C. porosus*. **In:** Wildlife management: Crocodiles and Alligators G. W. Webb; C. Manolis and P. J. Whitehead, pblsh. Surrey Beatty & sons Pty Limited.

Webb, G. J. W.; Manolis, S. C.; Dempsey, K. E. & Whitehead, P. J. (1987b). Crocodilian eggs: A functional overview. **In:** Wildlife management: Crocodiles and Alligators G. W. Webb; C. Manolis and

P. J. Whitehead, pblsh. Surrey Beatty & sons Pty Limited.

Whitehead, P. J.; Webb, G. J. W. & Seymour, R. S. (1990). Effect of incubation temperature on development of *Crocodylus johnstoni* embryos. *Physiological Zoology* **63**(5):949-964.

Yntema, C. L. & Mrsovsky, N. (1980). Sexual differentiation inhatchlings loggerheads turtles (Caretta caretta) incubated at different controlled temperatures. *Herpetologica* **36**:33-36

SIZE VARIATIONS IN NUTS, EGGS AND HATCHLINGS OF THE MARSH CAIMAN *CAIMAN CROCODILUS YACARE* INCUBATED IN NATURAL AND CONTROLLED CONDITIONS.

Rafaela Danielli Nicola[1] , Eliézer José Marques[2] , Konradin Metze[3]

[1] Master's Degree Course in Ecology and Conservation/CCBS, Federal University of Mato Grosso do Sul, P.O. Box 649, CEP 79070-900, Campo Grande, MS.

[2]Department of Biology/CCBS, Federal University of Mato Grosso do Sul.

[3]Department of Pathological Anatomy/FCM/UNICAMP

Summary:

We examined a sample of 17 *Caiman crocodilus yacare* nests in January 1998 in the Pantanal, Miranda and Abobral sub-regions. The average number of eggs per spawning was 25.93 + 5.37, with an approximate mass of 75.26 + 6.64 g, a length of 7.00 + 0.35 cm and a width of 4.31 + 0.13 cm. The size of the eggs varied between spawnings. The chicks hatched weighing an average of 49.16 + 5.00 g with an average total length of 23.51 + 1.12 cm, a head of 3.46 + 0.12 cm and a body of 11.74 + 0.51 cm. Chicks hatched in the laboratory had measurements equivalent to those hatched in natural conditions. Hatching rates differed between nests. In some spawnings, chicks were born with omphalitis (N = 14), only one of which died a few days later, while the others developed normally. **Keywords**: *Caiman crocodilus yacare*, nests, morphometry, Pantanal.

Introduction:

Reproductive investment is one of the main factors controlling the density of a population (Deitz & Hines, 1980). Considering the energy and time required to produce offspring, it is clear that this can be a costly activity (Gittleman & Thompson, 1988).

The idea that there is an energetic limitation to reproductive investment motivates a series of hypotheses about "trade off" relationships that have already been proven in various studies for different species.

In marine invertebrates, Sinervo & Edward (1988) observed important relationships between egg size and larval size and development in two species of urchins. McGinley *et al.* (1987) predicted a positive correlation between the amount of maternal resource (female size) and offspring size.

In species with iteroparous characteristics, Trivers (1972) suggests that the offspring of each reproductive event affects the production of offspring in subsequent events. Thus, there must be an optimum limit, maximizing reproductive investment, resulting from a trade-off between immediate fecundity and future reproductive potential (Charnov & Krebs, 1974).

Begon & Parker (1986) suggest that the size of the offspring of different females may vary due to the influence of clutch size. If competition between siblings is important for offspring survival, then larger females would produce larger eggs rather than more eggs.

In reptiles, the sources of energy that females have available to invest in reproduction are finite and can be analyzed through various factors, including: length of reproductive life; spawning frequency; spawning size; egg mass, number of offspring produced (Thorbjanarson, 1996; Bjorndal & Carr, 1989) or even energy expenditure on locomotion and nest building (Hays & Speakman, 1991).

Hays & Speakman (1991) observed that in sea turtles it is easier to quantify the time and energy spent on reproduction, since egg-laying takes place in discrete events and there is no parental care of eggs and hatchlings.

Although crocodilians have a more elaborate reproductive behavior (with parental care after spawning) Thorbjanarson (1996) suggests that the mass of the clutches is a good indicator of the female's reproductive investment.

The variations found in spawn sizes seem to be correlated to the size of the females, both for turtles (Hays & Speakman, 1991; Bjorndal & Carr, 1989; Hirth, 1980) and crocodilians (Thorbjanarson, 1996;

Campos & Magnuson, 1995 and Deitz & Hines, 1980). Thorbjanarson (1996) in his study showed that several species of crocodilians have a clear tendency to have spawn and egg sizes that depend on the size of the female.

Deitz & Hines (1980) point out that reproductive success depends not only on the number of eggs in the clutch, but also on the size of the eggs, since these determine the size of the young. This relationship (egg - offspring) was also considered by Sinervo & Huey (1990), when they carried out a manipulation experiment on lizards (*Sceloporus occidentals*), proving that motor performance is related to the size of the offspring.

In a study with *Caiman crocodilus yacare*, Campos & Magnuson (1995) found that variations in clutch size can be explained by the size of the females, but the size of the young and eggs is independent of the females.

As explained above, the amount of energy allocated to reproduction is limited by various factors, so energy investment in one reproductive aspect must interfere with the amount of energy left over for other aspects ("trade off"). Based on this concept, we carried out a study with the aim of verifying how aspects of reproductive investment in a population of *C. c. yacare* are interrelated. We sought to answer the following questions:

What are the clutches like and what is the hatching success for the same nesting season?

Regardless of the size of the female, is there a relationship between the number of eggs in a clutch and the size of the eggs?

Is there a family effect (genetic, characterized by the clutch) in the relationship between the size of eggs and chicks?

Material and methods:

Study area

The collections were made at the beginning of January 1998, the species' nesting season (Cintra, 1988; Crawshaw, 1987 and Campos, 1993), in the Pantanal sub-regions of Miranda and Abobral (Adâmoli, 1982), in the municipality of Corumbà, near the research base (BEP) of the Federal University of Mato Grosso do Sul, in areas of capoes and ridges of the Miranda and Vermelho rivers. We located and collected eggs from recent spawnings. We collected 14 spawns along with nest material, no more than five days old. The original position of the eggs was indicated with a pen mark and maintained during handling to prevent the embryo from detaching or rotating, which could lead to its death (Webb *et al.*, 1987). Two spawnings, after measuring the eggs, were not considered in the sample because they had too many eggs, suggesting multiple laying nests (Marques & Monteiro, 1995).

Methodology:

The 14 spawns were transported for incubation in the laboratory at the

Federal University of Mato Grosso do Sul, according to the methodology described in Chapter

III. The eggs from each spawning were distributed proportionally among the five different temperature treatments (approximately 5 eggs from each nest per treatment).

All the eggs were weighed on a semi-analytical scale and measured in width (smallest diameter) and length (largest diameter) using a 0.01 mm caliper.

The eggs were incubated in the laboratory (Chapter III) and after hatching each chick was immediately washed in running water and treated with Rifocin Spray (Rifamicin SV, Merrell Lepetit Farmacèutica LTDA) to prevent infections. A color code with colored beads was attached to the base of the caudal crests (Marques & Monteiro, 1995). All the chicks were weighed on a semi-analytical scale and measured using a 0.01 mm precision caliper for total length (TC - from the tip of the snout to the end of the tail), body (CP - from the tip of the snout to the back end of the cloaca) and head (CA - from the tip of the snout to the base of the skull).

The same measurements were also taken on chicks belonging to a spawning incubated under natural conditions, found in a riparian forest area of the Miranda River and transferred to the laboratory only at the end of the incubation period (approximately 15 days before hatching).

The data was analyzed using "Kolmogorov-Smirnov" normality tests, analysis of variance (ANOVA, and *a posteriori* SNK "Student-Newman-Keuls" test), multiple and simple linear regression and we also used "Box-Plot Whisker" graphs to illustrate the distribution of the dimensions evaluated for each spawning separately.

Result s:

A total of 17 spawns *of C. c. yacare* found in the riparian vegetation of the Miranda and Vermelho rivers were measured. Of these, 14 had eggs collected for incubation in the laboratory. Of the spawning sites found, two had very high numbers of eggs (49 and 46 eggs) and must have been multiple spawning nests, i.e. used by more than one female. Disregarding these spawnings, the average number of eggs per spawning was 25.93 + 5.37, the smallest being 16 and the largest 35 eggs.

Table 01 - Spawning size, dimensions and body mass of *Caiman crocodilus yacare* eggs and young.

		Average +dp	**Minimum**	**Maximum**
Spawning (n= 15)		25,93 + 5,37	16	35
EGGS n = 486	Mass (g)	75,26 + 6,64	44,53	92,57
	Length (cm)	7,00 + 0,35	4,35	8,16
	Width (cm)	4,31 + 0,13	3,66	4,80
CHILDREN (Inc. artificial) n = 254	Mass (g)	49,33 + 5,21	36,50	63,2
	CA (cm)	3,46 + 0,12	3,00	3,72
	CP (cm)	11,73 + 0,51	10,20	12,90
	CT (cm)	23,46 + 1,16	19,70	26,30
CHILDREN (Inc. natural) n = 30	Mass (g)	47,77 + 2,18	43,40	51,7
	CA (cm)	3,56 + 0,09	3,37	3,77
	CP (cm)	11,99 + 0,33	11,30	12,60
	CT (cm)	23,90 + 0,60	22,50	24,90
CHILDREN (Total) n = 284	Mass (g)	49,16 + 5,00	36,50	63,20
	CA (cm)	3,46 + 0,12	3,00	3,72
	CP (cm)	11,74 + 0,51	10,20	12,90
	CT (cm)	23,51 + 1,12	19,70	26,30

Average, minimum and maximum values for spawning size, eggs and chicks. CA - head;

CP - body; CT - total length.

According to Table 01, the eggs collected had an average mass of 75.26 + 6.64 g, an average length of 7.00 + 0.35 cm and an average width of 4.31 + 0.13 cm.

Considering all the chicks measured, the average body mass was 49.16 + 5 g and the average dimensions for head size (HC) were 3.58 + 1.92 cm, body size (BC) was 12.13 + 6.32 cm and total length (TL) was 23.51 + 1.25 cm. Analysis of variance showed that the chicks incubated in the laboratory had similar measurements to those incubated naturally (mass with $F_{,1284} = 2.602$ $p = 0.11$; body with $F_{1.284} = 0.016$ $p = 0.9$ and head with $F_{1.284} = 0.006$ $p = 0.938$). Only total length showed statistically significant differences between the two groups (total length with $F_{1,284} = 4.15$ $p = 0.04$).

Multiple linear Regressions carried out for each parameter measured in the cubs (head length with $R^2 = 18.6\%$, body length with $R^2 = 11.2\%$, total length with $R^2 = 16.8\%$ and mass with $R^2 = 28.8\%$) and the measurements taken for the eggs (egg mass, length and width), showed that the dimensions of the alligators at birth are not very dependent on the dimensions of the corresponding eggs. In the specific case of the mass of the hatchlings, which showed the highest value for R^2 , it was observed that this was more closely related to the mass of the egg ($R^2 = 27\%$) and, to a lesser extent, to the length (R2 = 13%).

The average hatch rate between the different spawns was 68.30 + 13.0%. Only one nest had a very low rate of only 31.8%. Excluding this nest, the average hatch rate was 71.10 + 7.90%, ranging from a minimum of 60.0% to a maximum of 82.9%. Of the eggs incubated, 14 chicks hatched with an open yolk (Fig. 01).

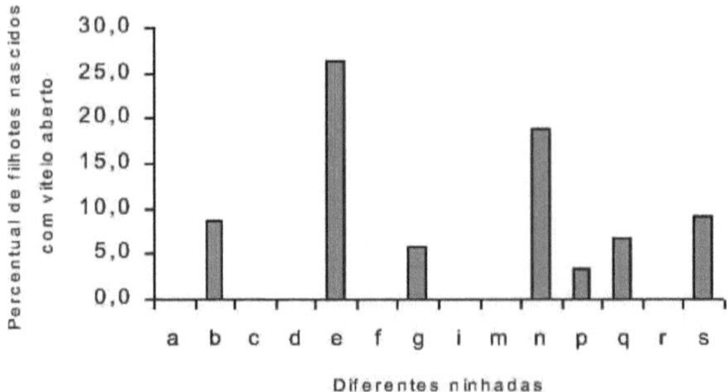

Figure 01 - Percentage of pups born with open calves in different litters (pups with exposed calves/total pups born in litter * 100).

These statistical tests are based on the assumption that the elements are independent of each other, but eggs from the same clutch generally have particular characteristics and should be analyzed individually.

Of the 17 spawns studied, 14 were incubated under the same laboratory conditions, allowing for the results described below. Using the Kolmogorov-Smirnov test for continuous variables, we verified the normality of the data for spawning size (p=0.91), number of chicks born (p=0.99) and hatching rate (p=0.80). The correlation between spawning size and hatching rate was not significant (R^2 = 21%), while the number of chicks hatched in each spawning was highly dependent (R^2 = 77%) on the number of eggs.

The analysis of variance between the spawnings for each dimension taken, both in eggs and chicks, indicated a statistically significant difference ($p < 0.001$). The "Student-Newman-Keuls" *a posteriori* test showed that the spawns differed more in terms of the size of the eggs than in terms of the chicks. This test also showed that one particular spawning (with 33 eggs) differed significantly from the others for all the variables evaluated.

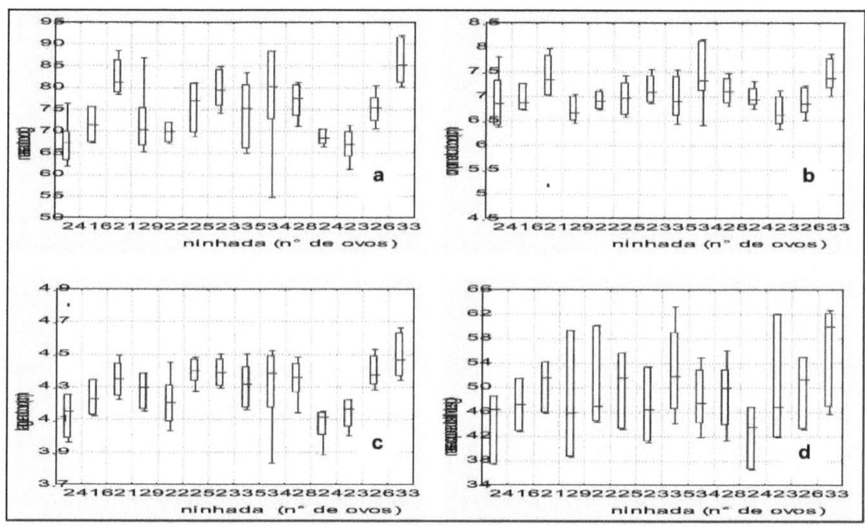

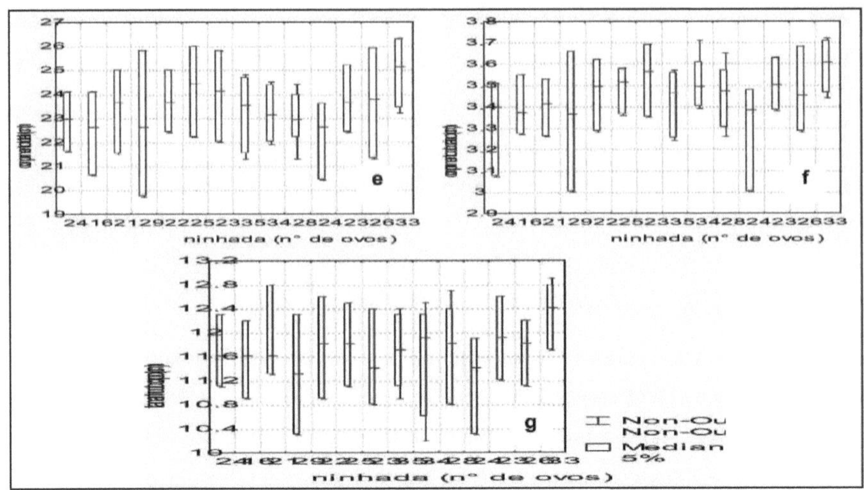

Figure 02 - Distribution of measurements taken for eggs and chicks in relation to each clutch. **a** - egg mass; **b** - egg length; **c** - egg width; **d** - body mass of chicks; **e** - total length (TL); f - skull length (SL); **g** - body size (BS).

The graphs in Fig. 02 show the distribution of median values for the measurements of eggs and chicks in each spawning.

Discussion:

The values found for the mass, length and width of all the eggs and for the measurements of the chicks in this study were similar to those published independently by Crawshaw & Schaller (1980) and Cintra (1988) for the same species. Despite the small number of nests analyzed, the hatching rate of the clutches had a normal distribution and the data showed that a greater number of eggs in the nest does not favor hatching success, since the relationship between hatching rate and nest size is not significant ($R^2 = 0.20$, n =14 $p < 0.05$).

Deitz & Hines (1980) found a strong correlation between egg mass and offspring for *Alligator mississippiensis* ($R^2 = 90\%$). Our results showed that there is a significant ($p<0.001$) but weak positive relationship between the dimensions of all eggs and chicks. Comparing the average values, for each clutch, between egg mass and total length (TL) through linear regression, we observed that the variations in TL are explained by only 38% (n = 14, $p < 0.001$). For the whole sample, egg mass is the factor that best explains variations in chick dimensions, influencing their body mass the most ($R^2 = 27\%$). In a study carried out by Campos & Magnuson (1995), the authors found that 52% of the variation in the average length of chicks per clutch *of C. c. yacare* was explained by the average variation in egg mass.

The results of this study indicate that clutch size is not correlated with any egg measurement (R2 = 14% for mass, 13% for width and 7% for length with $p< 0.001$ and n = 368). Although studies indicate that larger spawn tend to have smaller eggs (Begon & Parker, 1986), several studies carried out with turtles and crocodilians indicate that the relationship between egg and nest size is not always clear. These parameters seem to be more directly related to the size of the female (Campos & Magnuson, 1995; Thorbjanarson, 1996; Bjorndal & Carr, 1989; Deitz & Hines, 1980). Bjorndal & Carr (1989) point out that it is difficult to say exactly how strong this correlation is.

It was not possible to obtain information on the size of the females, as they were not found near the nests at the time of collection. Campos & Magnuson (1995) state that 67% of the variation in the number of eggs per pantanal caiman nest is explained by the size of the females.

Crawshaw & Schaller (1980) reported a considerable variation in the average size of eggs from different clutches and the results presented here contribute to this information, indicating that both eggs and chicks from each clutch have significantly different measurements (Anova $p<0.001$).

Thorbjanarson (1996), working with various crocodilian species, observed that the differences between nests can be explained by a "family factor" (parental characteristics), measured through the dimensions of the female. The graphs in Figure 02 show that the distribution of the parameters measured for eggs and chicks show different patterns for each brood, which can be seen in the different shifts of the median for the same variable between the nests. These data are consistent with the hypothesis of a "family factor" or brood factor, which particularizes the characteristics of each nest.

Bjorndal & Carr (1989) showed that the coefficient of variation for eggs was smaller than that for nests (3.7 and 18.7 respectively) for *Chelonia mydas* . Table 02 shows that for the marsh caiman, the variations between nest sizes were also greater than the variations for egg and offspring sizes (coefficients of variation).

Table 02 - Values of the coefficients of variation for spawning size, eggs and young of *Caiman crocodilus yacare*, calculated from the data in Table 01.

Coefficient of variation		
DESOVAS (n= 15)		0,207
EGGS n = 486	Pasta	0,08

	Length	0,05
	Width	0,03
	Pasta	0,101
CHILDREN	CA	0,035
(Total) n = 284	CP	0,043
	CT	0,047

CA - head; CP - body; CT - total length.

Analysis of variance between chicks hatched in natural conditions and chicks hatched in the laboratory showed a high probability of similarity between the groups, with only the total length showing statistically significant differences, but with a **p-value** close to the maximum acceptable limit (p=0.04), indicating that the laboratory conditions did not interfere with their dimensions. The interpretation of this result requires caution, since our laboratory sample consisted of 14 different broods, while the field sample consisted of only 1 nest.

Of the eggs incubated, 14 chicks hatched with an open yolk sac, an alteration indicated by Matushima & Ramos (1995) as omphalitis. The incubation time for the cases mentioned here was very long, exceeding 80 days, and the reactions of the chicks at hatching differed from the others. The animals didn't vocalize and had difficulty breaking through the shell, having to be helped by hand. Of these, only one died a few days after hatching; in the others, the yolk was reabsorbed after approximately 7 days. These animals came from 5 different spawnings, one of which gave rise to 5 animals with omphalitis. Considering that, for the artificial incubation process, the distribution of the eggs in the different treatments was carried out maintaining the proportionality of their origin, we conclude that the omphalitis detected in these chicks was not induced by the incubation process, but must be related to the clutch (Fig. 01).

The development of all the puppies was monitored for 14 months. In the case of those born with omphalitis, development did not differ from the others, and it was only possible to identify them by color coding.

Bibliographical references:

Adâmoli, J. (1982). The Pantanal and its phytogeographic relations with the Cerrados. XXXII National Botanical Congress Proceedings:109-119.

Begon, M. 7 Parker, G. A. (1986). Should egg size decrease with age? Oikos **47**: 293-302.

Bjorndal, K. A. & Carr, A. (1989) Variation in clutch size and egg size in the green turtle nesting population at Tortuguero, Costa Rica. Herpetologica, **45**(2)181-189.

Campos, Z. & Magnuson, W. (1995).Relationships between rainfall, nesting habitat and fecundity of Caiman crocodilus yacare in the Pantanal, Brazil. *J. of Trop. Ecol.* **11**:351-358.

Campos, Z. (1993). Effect of habitat on survival of eggs and sex ratios of hatchlings *of Caiman crocodilus yacare* in the Pantanal, Brazil. *J. Herpetology* **27**(2):127-132.

Charnov, E. L. & Krebs, J. R. (1974). On clutch size and fitness. Ibis, **116**:217-219.

Cintra, R. (1988). Nesting ecology of the Paraguayan Caiman (Caiman yacare) in the Brazilian Pantanal. *J. Herpetology,* **22**(2): 219-222.

Crawshaw, P. G. & Schaller, G. B. (1980). Nesting of the Paraguayan Caiman (Caiman yacare) in Brazil. *Papéis Avulsos de Zoologia*, **33** (18):283-292.

Crawshaw, P.G. (1987). Nesting ecology of the Paraguayan (*Caiman yacare*) in the Pantanal of Mato Grosso, Brazil. MS Thesis (unpubl.). Gainesville, University of Florida, 68pp.

Deitz, D. C. & Hines, T. C. (1980). Alligator Nesting in North-Central Florida. Copeia **2**:249-258.

Gittleman, J. L. & Thompson, S. D. (1988). Energy allocation in mammalian reproduction. American Zoologist, **20**:507-523.

Hays, G. C. & Speakman, J. R. (1991) Reproduction investment and optimal clutch size of loggerhead sea turtles (*Caretta caretta*). J. Animal Ecology, **60**:455-462.

Hirth, H. F. (1980). Some aspects of the nesting behavior and reproductive biology of sea turtles. American Zoology, **20**:507-523.

Marques, E. J. & Monteiro, E. L. (1995). Ranching of Caiman crocodilus yacare in the Pantanal of Mato Grosso do Sul, Brazil. **In:** Conservation and Management of Caimans and Cocodiles of Latin America. Alejandro Larriera and Luciano M. Verdade. Published by Fundacion Banco Bica, Santo Tomé, Santa Fe, Argentina.

Matushima, E. R. & Ramos, M. C. C. (1995). Some pathologies in alligator breeding in Brazil. **In:** La conservacion y el manejo de caimanes y cocodrilos de America Latina. Alejandro Larriera and Luciano M. Verdade. Published by Fundacion Banco Bica, Santo Tomé, Santa Fe, Argentina.McGinley, M. A. (1989). The influence of a positive correlation between clutch size and offspring fitness on the optimal offspring size. Evolutionary Ecology, **3**:150-156.

McGinley, M. A.; Temme, D. H. & Geber, M. A. (1987). Parental investment in variable environments: theoretical and empirical considerations. Amer. Natur. **130**:370-398.

Sinervo, B. & Huey, R. B. (1990). Allometric engineering: anexperimental test of the causes of interpopulation differences in performance. *Science* **248:**1106-1109.

Sinervo, B. & McEdward, L. R. (1988). Developmental consequenses of an evolutionary change in egg size:an experimental test. *Evolution*, **42** (5):885-899.

Thorbjanarson, J. B. (1996). Reproductive characteristics of the order Crocodylia. Herpetologica, **52**(1):8-24.

Trivers, R. L. (1972). Parental investment and sexual selection. **In** Sexual selection and the Descent of Man, 1871-1971.(ed. By B. Campbell) pp136-179. Aldine Press, Chicago.

Webb, G. J. W.; Manolis, S. C.; Dempsey K. E. & Whitehead P. J. (1987). Crocodilian eggs: A functional overview. **In:** Wildlife management: Crocodiles and Alligators G. W. Webb; C. Manolis and P. J. Whitehead, pblsh. Surrey Beatty & sons Pty Limited.

GENERAL DISCUSSION

This work, made up of four independent but interconnected parts, sought to elucidate issues relating to the processes of sex determination and nesting in the marsh caiman (*C. c. yacare*). The fieldwork, involving observations of nesting, made it possible to design the experiment presented in Chapter I, in which we analyzed the thermal behavior of a forest nest under natural conditions. The results of this work were used in the theoretical and experimental study presented in Chapter III.

The development of the incubation system, described in Chapter II, allowed us to obtain suitable and safe laboratory conditions for incubating the eggs at the desired temperatures, making it possible to carry out the experiments presented in Chapters III and IV.

In the third chapter we use a model to analyze the influence of thermal fluctuations on the sex determination of the species. Identifying the sex of the chicks through histological analysis of the gonads proved to be the most accurate technique for obtaining this data.

During the course of this work, we encountered some difficulties which, within the existing working conditions, were overcome. For example, the issue of sex identification of alligators, which initially seemed simple, required a relatively large amount of work, including involving other research institutions. The solution found was through a histological study of the gonads, associating it with the characteristics of the external genitalia.

Maintaining the samples in captivity also required special dedication, as the conditions available were not ideal for this type of pioneering experiment. Despite the efforts and financial support given to the experiment, over the period of keeping the animals, many developed a pathology of as yet unknown cause and died. The symptoms observed were recorded and material was collected for future analysis.

The interpretation of the results presented in Chapter III was limited to the amount of information in the literature, representing a difficulty that so far cannot be overcome.

As a final contribution to the study of the marsh caiman, the four papers presented offer a current bibliographic list that will certainly be useful for continuing this line of research.

BIBLIOGRAPHICAL REFERENCES

Adâmoll, A. (1986). The dynamics of flooding in the Pantanal. pp. 51-56. **In**: Proceedings of the 1st Symposium on the Natural and Socio-Economic Resources of the Pantanal. EMBRAPA-DDT, Brasilia, Brazil.

Boggiani, P. C. and Coimbra, A. M. (1996). The plains and the wetlands. **In** P.T.Z. Antas and I. L. S. Nascimento, ed. Tuiuiù - Sob os céus do Pantanal - Biologia e conservaçâo do tuiuiù (*Jabiru mycteria*). Empresa de Artes, Sao Paulo pp. 18-23.

Brazaits, P.; C. Yamashita & G. Rebelo. (1990) A summary report of the CITES Central South America caiman study: phase I: Brazil. pp.100-115. In Proc. 9th Working Meeting of the IUCN/SSC/Crocodile Specialist Group, Papyua-New Guinea. **Vol. I IUCN** Pub. N. S. Gland, Switzerland.

Bull, J. J. (1983). Evolution of sex determining mechanisms. Chapter 9. Edited by The Benjamin/Cummings pbsh. Co. Menlo Park, California.

Cadavid Garcia, E. A. (1984). The climate of the Mato Grosso Pantanal. EMBRAPA-UEPAE Corumbâ. Technical Circular **14**. 37 pp.

Campos, Z. & Magnuson, W. (1995). Relationships between rainfall, nesting habitat and fecundity of *Caiman crocodilus yacare* in the Pantanal, Brazil. *J. of Trop. Ecol.* **11**:351-358.

Charnier, M. (1966). Action de la temperature sur la sex-ratio chez l'embryon d'*Agama agama* (Agamidae, Lacertilien). *Soc. Biol. Quest. Afr.* **160**:620-622.

Cohen, M. M. & Gans, C. (1970). The chromosomes of the Order Crocodilia. *Cytogenetcs* **9**:81105.

Crawshaw, P.G. 1987. Nesting ecology of the Paraguayan caiman (*Caiman yacare*) in the Pantanal of Mato Grosso, Brazil. MS Thesis (unpubl.). Gainesville, University of Florida, 68pp.

Ferguson, M. W. J & JOANEN, T. (1982). Temperature of egg incubation determines sex in *Alligator mississippiensis. Nature*, **296**:850-854.

Groombridge, B. (1987). Distribution and status of world crocodilians. pp 9-21 . **In** Wildlife management: Crocodiles and Alligators G. W. Webb; C. Manolis and P. J. Whitehead, pblsh. Surrey Beatty & sons Pty Limited.

Hutton, J. M. (1987). Incubation temperature, sex ratios and sex determination in a population of Nile crocodiles (*Crocodylus niloticus). J. Zool. Lond.* **211**: 143-155.

Lang, J. W.; Andrews, H. & Whitaker, R. (1989). Sex determination and sex ratios in *Crocodilus palustris. Amer. Zool.*, **29**:935-952.

Marques, E. J. & Monteiro, E. L. (1995). Ranching *of Caiman crocodilus yacare* in the Pantanal. pp. 189-211. **In**: Conservation and Management of Caimans and Crocodiles of Latin America. Edited by Alejandro Larriera & Luciano M. Verdade Publ. Fundación Banco bica, Santo Tomé, Santa Fe,

Argentina.

Mrosovsky, N.; Dutton, P. H. & Whitemore, C. P. (1984). Sex ratios of two species of sea turtles nesting in Suriname. *Can.J. Zool.* **62**:2227-2239.

Pough, F. H.; Andrews, R. M.; Cadle J. E.; Crump, M. L.; Savitzky, A. H. & Wells, K. D. (1998). Herpetology. Chapter 7 Prentice Hall, United States of America, 577pp.

RADAMBRASIL (1982a). Survey of natural resources. Vol. **27**: Sheet SE. 21 Corumbâ.

RADAMBRASIL (1982b). Survey of natural resources. Vol. **28**: Sheet SF. 21 Campo Grande.

Webb, G. J. W.; Manolis, S. C.; Dempsey, K. E. & Whitehead, P. J. (1987). Crocodilian eggs: A functional overview. **In:** Wildlife management: Crocodiles and Alligators G. W. Webb; C. Manolis and P. J. Whitehead, pblsh. Surrey Beatty & sons Pty Limited.

Whitehead, P. J.; Webb, G. J. W. & Seymour, R. S. (1990). Effect of incubation temperature on development of *Crocodylus johnstoni* embryos. *Physiological Zoology* **63**(5):949-964.

Yntema, C. L. & Mrosovsky, N. (1980). Sexual differentiation in hatchling loggerhead (*Caretta caretta*) incubated at different controlled temperatures. *Herpetologica*, **36**:33-36.

Printed by Books on Demand GmbH, Norderstedt / Germany